Math Magicians: Functional Analysis in Quantum Field Theory All in One Skills Practice Workbook

Jamie Flux

https://www.linkedin.com/company/golden-dawn-engineering/

Collaborate with Us!

Have an innovative business idea or a project you'd like to collaborate on?
We're always eager to explore new opportunities for growth and partnership.
Please feel free to reach out to us at:

https://www.linkedin.com/company/golden-dawn-engineering/

We look forward to hearing from you!

Contents

1 Introduction to Functional Analysis   **13**

Overview of Functional Analysis 13

Importance in Mathematical Physics 13

Basic Concepts 13

1   Vector Spaces 13

2   Norms 13

3   Convergence 14

Practice Problems 1 14

Answers 1 15

Practice Problems 2 16

Answers 2 17

Practice Problems 3 18

Answers 3 19

2 Normed Spaces and Banach Spaces   **21**

Normed Vector Spaces 21

Completeness and Cauchy Sequences 21

Banach Spaces 21

Applications in Quantum Theory 22

Examples Relevant to Quantum Theory 22

Practice Problems 1 22

Answers 1 23

Practice Problems 2 24

Answers 2 25

Practice Problems 3 27

Answers 3 27

3 Hilbert Spaces and Inner Product Spaces   **29**

Inner Product Spaces 29

Properties of Hilbert Spaces 29

1   Examples 29

Orthonormal Bases 30

Projection Theorems 30

1   Orthogonal Projection 30

2   Best Approximation Theorem 30

Algorithms for Orthogonalization 30

Practice Problems 1 30

Answers 1 32

Practice Problems 2 32

Answers 2 34

Practice Problems 3 36

Answers 3 37

4 Linear Operators on Hilbert Spaces   **40**

    Bounded and Unbounded Linear Operators 40

    1   Examples of Bounded Operators 40

    2   Examples of Unbounded Operators 40

    Adjoint Operators 40

    1   Properties of Adjoint Operators 40

    Operator Norms and Their Properties 41

    Study of Properties in Hilbert Spaces 41

    1   Spectral Properties 41

    2   Domain and Range Properties 41

    3   Algorithm for Operator Norm Computation 41

    Practice Problems 1 41

    Answers 1 42

    Practice Problems 2 43

    Answers 2 44

    Practice Problems 3 45

    Answers 3 46

5 Spectral Theorem for Bounded Operators   **48**

    Introduction to the Spectral Theorem 48

    Bounded Self-Adjoint Operators 48

    Spectral Measures and Projection-Valued Measures 48

    1   Properties of Spectral Measures 48

    Applications in Quantum Mechanics 49

    Integral Representation of Operators 49

    1   Functional Calculus 49

    Key Examples 49

    1   Example: Position Operator 50

    Summary of Fundamental Results 50

    Practice Problems 1 50

    Answers 1 51

    Practice Problems 2 52

    Answers 2 53

    Practice Problems 3 54

    Answers 3 55

6 Self-Adjoint and Unbounded Operators   **57**

    Unbounded Operators on Hilbert Spaces 57

    Self-Adjointness and Symmetric Operators 57

    Deficiency Indices and Their Role 57

    Spectral Theorem for Unbounded Operators 58

    Von Neumann's Criteria for Self-Adjointness 58

    Applications in Quantum Field Theory 58

    Practice Problems 1 58

    Answers 1 60

    Practice Problems 2 60

    Answers 2 62

    Practice Problems 3 63

    Answers 3 64

7   Distributions and Generalized Functions   **65**

    Introduction to Distributions 65

    1   Test Functions 65

    2   Definitions and Basic Properties 65

    The Dirac Delta Function 65

    Operations on Distributions 66

    1   Differentiation of Distributions 66

    2   Addition and Scalar Multiplication 66

    3   Convolution of Distributions 66

    Examples and Applications 66

    1   Fundamental Examples 66

    Practice Problems 1 67

    Answers 1 68

    Practice Problems 2 70

    Answers 2 71

    Practice Problems 3 73

    Answers 3 74

8   Fourier Analysis in Functional Spaces   **76**

    Introduction to Fourier Transform in L^2 Spaces 76

    1   Definition of the Fourier Transform 76

    2   Properties of the Fourier Transform 76

    3   Inversion Formula 76

    Plancherel Theorem 77

    1   Statement of Plancherel Theorem 77

    2   Applications of Plancherel Theorem 77

    3   Proof of Plancherel Theorem 77

    Applications to Solving Differential Equations 77

    1   Heat Equation 77

    2   Wave Equation 77

    Convolution Theorem 78

    1   Convolution and Fourier Transform 78

    Practice Problems 1 78

    Answers 1 80

    Practice Problems 2 81

    Answers 2 83

    Practice Problems 3 84

    Answers 3 85

9   Sobolev Spaces   **87**

    Introduction to Sobolev Spaces 87

    Weak Derivatives 87

    Sobolev Space $W^{1,p}(\Omega)$ 87

    Embedding Theorems 87

    Poincaré Inequality 88

    Extension Theorems 88

    Sobolev Inequality 88

    Significance in Field Equations 88

    Practice Problems 1 88

    Answers 1 90

    Practice Problems 2 91

    Answers 2 92

    Practice Problems 3 95

    Answers 3 96

10 Rigged Hilbert Spaces and Gelfand Triplets   98

Introduction to Rigged Hilbert Spaces 98
Construction of Gelfand Triplets 98
Role in Spectral Decomposition of Unbounded Operators 98
1   The Spectral Theorem 98
2   Applications to Quantum Mechanics 99
3   Example: The Harmonic Oscillator 99
Practice Problems 1 99
Answers 1 100
Practice Problems 2 102
Answers 2 103
Practice Problems 3 104
Answers 3 104

11 Introduction to Quantum Mechanics in Hilbert Space   106

Quantum States as Vectors 106
1   Superposition Principle 106
2   Orthogonality and Orthonormality 106
Operators as Observables 106
1   Eigenvalues and Eigenvectors 106
2   Projection Operators 107
Probabilistic Interpretation 107
1   Born Rule 107
2   Expectation Values 107
State Evolution 107
1   Schrödinger Equation 107
2   Unitary Evolution 107
Measurement Postulate 107
Example: Spin-$\frac{1}{2}$ Systems 108
1   Spin Operators 108
2   Measurement and Probabilities 108
Practice Problems 1 108
Answers 1 109
Practice Problems 2 110
Answers 2 112
Practice Problems 3 113
Answers 3 114

12 Quantum Observables and Commutation Relations   117

Quantum Observables as Operators 117
Eigenvalues and Eigenvectors of Observables 117
Commutation Relations 117
Canonical Commutation Relations 117
The Uncertainty Principle 118
Expectation Values and Variance 118
Example: Position and Momentum 118
Operator Algebra and Quantum Dynamics 118
Practice Problems 1 119
Answers 1 120
Practice Problems 2 121
Answers 2 122
Practice Problems 3 124
Answers 3 125

13 Quantum Systems with Infinite Degrees of Freedom   **127**

Introduction to Infinite Degrees of Freedom 127

The Harmonic Oscillator as a Model 127

Field Theory: Introducing Continuity 127

Quantum Fields 127

Canonical Quantization 128

Basis for Quantum Field Theory 128

Spectral Theory in Infinite Dimensional Spaces 128

Mathematical Challenges 128

Transition to Quantum Field Theory 128

Practice Problems 1 129

Answers 1 130

Practice Problems 2 131

Answers 2 132

Practice Problems 3 134

Answers 3 135

14 Canonical Quantization of Fields   **137**

Classical Field Theory 137

Canonical Quantization Process 137

Field Operators and Their Commutation Relations 137

Creation and Annihilation Operators 137

Quantization Example: Real Scalar Field 138

Role of Commutation Relations 138

Formulation in Fock Space 138

Practice Problems 1 138

Answers 1 140

Practice Problems 2 141

Answers 2 142

Practice Problems 3 144

Answers 3 145

15 Boson Fields in Hilbert Space Framework   **148**

Introduction to Bosonic Fields 148

Quantization of Scalar Boson Fields 148

Creation and Annihilation Operators 148

Fock Space Construction 149

Connection to Quantum Fields: Scalar and Vector Fields 149

Applications in Scalar and Vector Fields 149

Practice Problems 1 149

Answers 1 150

Practice Problems 2 152

Answers 2 152

Practice Problems 3 153

Answers 3 155

16 Fermion Fields and Anticommutation Relations   **157**

Introduction to Fermionic Fields 157

Anticommutation Relations 157

Quantization of Fermionic Fields 157

Fermi-Dirac Fock Space 157

Applications of Anticommutation Relations 158

Practice Problems 1 158

Answers 1 160

Practice Problems 2 161
Answers 2 162
Practice Problems 3 163
Answers 3 164

17 Spectral Theory in Quantum Field Theory **166**
Introduction to Spectral Theory 166
Field Operators in Quantum Field Theory 166
Spectral Theorem for Self-Adjoint Operators 166
Energy Spectrum Analysis 166
Applications to Particle Physics 167
Implications for Quantum Field Theory 167
Spectral Representation in Quantum Field Theory 167
Practice Problems 1 167
Answers 1 169
Practice Problems 2 170
Answers 2 170
Practice Problems 3 171
Answers 3 172

18 Scattering Theory and the S-Matrix **174**
Introduction to Scattering Theory 174
Asymptotic States 174
The S-Matrix 174
Lippmann-Schwinger Equation 175
Mathematical Formulation of Scattering Processes 175
Applications of the S-Matrix 175
Practice Problems 1 175
Answers 1 176
Practice Problems 2 178
Answers 2 179
Practice Problems 3 180
Answers 3 181

19 Wightman Axioms and Mathematical Foundations of QFT **183**
Introduction to Wightman Axioms 183
Axiom I: Existence of Hilbert Space 183
Axiom II: Field Operators 183
Axiom III: Covariance 183
Axiom IV: Spectrum Condition 184
Axiom V: Uniqueness of the Vacuum 184
Construction of Quantum Fields 184
Ensuring Mathematical Rigor in QFT 184
Practice Problems 1 184
Answers 1 186
Practice Problems 2 187
Answers 2 188
Practice Problems 3 189
Answers 3 191

20 Haag-Ruelle Theory   193

Foundations of Haag-Ruelle Theory 193
Asymptotic Completeness 193
Haag's Theorem 193
Ruelle's Formulation 194
Roles in Quantum Field Theory 194
Practice Problems 1 194
Answers 1 195
Practice Problems 2 197
Answers 2 198
Practice Problems 3 199
Answers 3 200

21 Algebraic Quantum Field Theory   202

Introduction to Algebraic Methods 202
C*-Algebras in Quantum Field Theory 202
1   Examples of C*-Algebras 202
Von Neumann Algebras 202
1   Properties and Examples 202
Applications in Quantum Field Theory 203
1   Local Algebras and Quantum Fields 203
2   State Spaces and Representations 203
Implementing Algebraic Methods 203
Practice Problems 1 203
Answers 1 204
Practice Problems 2 206
Answers 2 207
Practice Problems 3 208
Answers 3 209

22 Operator Product Expansion and Microcausality   212

Introduction to Operator Product Expansion 212
Convergence of the Operator Product Expansion 212
Microcausality in Field Theory 212
1   Implications of Microcausality on OPE 213
Algorithmic Approach in OPE Calculation 213
Examples of OPE in Quantum Field Theory 213
Practice Problems 1 213
Answers 1 215
Practice Problems 2 216
Answers 2 217
Practice Problems 3 218
Answers 3 219

23 Renormalization and Functional Analysis   221

Introduction to Renormalization in Quantum Field Theory 221
Regularization Techniques 221
1   Dimensional Regularization 221
2   Pauli-Villars Regularization 221
Renormalization Group Equations 222
Functional Analysis in Renormalization 222
1   Banach and Hilbert Spaces 222
2   Operator Theory 222
Application of Functional Analysis in Regularization and Renormalization 222

1	Spectral Theory in Field Regularization	222
2	Constructive Field Theory	222
	Practice Problems 1	223
	Answers 1	224
	Practice Problems 2	226
	Answers 2	227
	Practice Problems 3	229
	Answers 3	230

24 Functional Integrals and Path Integrals — **233**

	Introduction to Functional Integration	233
	Path Integral Formulation of Quantum Mechanics	233
1	Derivation from Schrödinger's Equation	233
	Path Integral Formulation in Quantum Field Theory	234
1	Generating Functionals and Correlation Functions	234
	Feynman Path Integrals: A Computational Approach	234
	Functional Methods in Statistical Field Theory	235
1	Connection to Classical Statistical Mechanics	235
	Practice Problems 1	235
	Answers 1	237
	Practice Problems 2	238
	Answers 2	240
	Practice Problems 3	241
	Answers 3	243

25 Gauge Theories and BRST Symmetry — **245**

	Gauge Invariance	245
	Quantization of Gauge Fields	245
	BRST Formalism	246
1	BRST Charge and Cohomology	246
2	Applications and Implications	246
	Algorithm for Gauge Fixing and BRST Quantization	246
	Examples of Gauge Theories	246
	Practice Problems 1	247
	Answers 1	248
	Practice Problems 2	250
	Answers 2	251
	Practice Problems 3	252
	Answers 3	253

26 Index Theorems and Applications in QFT — **255**

	Introduction to Index Theorems	255
	Proof Outline of the Atiyah-Singer Index Theorem	255
1	Elliptic Complexes and Symbols	255
2	Topological K-Theory	255
3	Heat Equation Approach	256
	Applications to Anomalies in Quantum Field Theory	256
1	Chiral Anomalies	256
2	Gravitational Anomalies	256
	Algorithmic Approach to Index Calculation	256
	Practice Problems 1	256
	Answers 1	258
	Practice Problems 2	259
	Answers 2	259

Practice Problems 3 260
Answers 3 261

27 Anomalies in Quantum Field Theory   263
Introduction to Anomalies 263
Mathematical Origin of Anomalies 263
1   Chiral Anomalies 263
2   Gravitational Anomalies 263
Implications for Physical Theories 264
1   Cancellation of Anomalies 264
2   Applications to the Standard Model 264
Algorithmic Approach to Detect Anomalies 264
Practice Problems 1 264
Answers 1 265
Practice Problems 2 266
Answers 2 267
Practice Problems 3 268
Answers 3 270

28 Non-Perturbative Methods in Functional Analysis   271
Introduction to Non-Perturbative Techniques 271
Operator Algebras in Quantum Field Theory 271
1   C*-Algebras 271
2   Von Neumann Algebras 271
Topological Methods in Quantum Field Theory 271
1   Fiber Bundles 272
2   Homotopy and Cohomology 272
Applications in Small-Coupling and Strong-Coupling Regimes 272
1   Small-Coupling Regimes 272
2   Strong-Coupling Regimes 272
Algorithm for Non-Perturbative Analysis 273
Practice Problems 1 273
Answers 1 274
Practice Problems 2 276
Answers 2 277
Practice Problems 3 278
Answers 3 279

29 Thermal Field Theory and KMS Condition   280
Introduction to Thermal Quantum Field Theory 280
Statistical Mechanics in Quantum Field Theory 280
The KMS Condition 280
1   Implications of the KMS Condition 280
Thermal Quantum Field Theory Framework 281
1   Path Integral Formulation 281
2   Thermal Propagators 281
3   Fermionic Systems 281
Algorithm for Calculating Thermal Green's Functions 281
Practice Problems 1 282
Answers 1 283
Practice Problems 2 284
Answers 2 285
Practice Problems 3 287
Answers 3 288

30 Quantum Fields on Curved Spacetimes   **290**

Introduction to Quantum Fields and Curved Spacetimes 290

Mathematical Foundations 290

1   The Covariant Derivative 290

Field Quantization in Curved Spacetimes 291

1   Mode Decomposition 291

2   Curved Spacetime Effects 291

Examples of Quantum Fields in Curved Spacetimes 291

1   Schwarzschild Spacetime 291

2   Friedmann-Lemaître-Robertson-Walker (FLRW) Universe 291

Practice Problems 1 292

Answers 1 293

Practice Problems 2 294

Answers 2 296

Practice Problems 3 297

Answers 3 299

31 Supersymmetry and Super Hilbert Spaces   **301**

Introduction to Supersymmetric Quantum Field Theories 301

Construction of Super Hilbert Spaces 301

1   Supervectors and Superoperators 301

2   Inner Product and Norm in Super Hilbert Spaces 302

Applications of Supersymmetry in Quantum Field Theory 302

Role of Super Hilbert Spaces in Supersymmetric Field Theories 302

Practice Problems 1 302

Answers 1 303

Practice Problems 2 304

Answers 2 305

Practice Problems 3 306

Answers 3 308

32 Noncommutative Geometry in Quantum Field Theory   **310**

Introduction to Noncommutative Geometry 310

Algebraic Framework 310

1   Spectral Triples 310

Noncommutative Spaces in Quantum Field Theory 311

1   Moyal Product 311

2   Field Theory on Noncommutative Spaces 311

Applications and Implications 311

Practice Problems 1 311

Answers 1 313

Practice Problems 2 314

Answers 2 315

Practice Problems 3 316

Answers 3 317

33 Microlocal Analysis in Quantum Field Theory   **320**

Introduction to Microlocal Analysis 320

Wavefront Sets and Singularities 320

1   Definition and Properties 320

Propagation of Singularities 320

1   Microlocal Propagation Theorem 321

2   Applications in Quantum Field Theory 321

Algorithms for Identifying Singularities 321

1   Implementing Hamiltonian Flow 321
Practice Problems 1 321
Answers 1 323
Practice Problems 2 324
Answers 2 325
Practice Problems 3 326
Answers 3 327

Chapter 1

Introduction to Functional Analysis

Overview of Functional Analysis

Functional analysis is a branch of mathematical analysis that studies spaces of functions and the functional operations on them. It provides the underpinning for various theories in mathematics and physics, particularly in the realm of quantum mechanics and quantum field theory. The methods of functional analysis have become essential in understanding the structure of infinite-dimensional spaces, where classical intuition often fails.

Importance in Mathematical Physics

The importance of functional analysis in mathematical physics lies in its ability to provide a rigorous framework for discussing infinite-dimensional vector spaces that arise naturally in the formulation of physical theories. For example, in quantum mechanics, states of a physical system are represented by vectors in a Hilbert space, and observables are linear operators on these spaces. Functional analysis bridges the gap between abstract algebraic formalism and the measurable entities in physical experiments.

Basic Concepts

1   Vector Spaces

A vector space V over a field $\mathbb{F}$ is a set equipped with an addition operation $+ : V \times V \to V$ and a scalar multiplication operation $\cdot : \mathbb{F} \times V \to V$, satisfying the following axioms for all $\mathbf{u}, \mathbf{v}, \mathbf{w} \in V$ and $a, b \in \mathbb{F}$:

1. $\mathbf{u} + \mathbf{v} = \mathbf{v} + \mathbf{u}$ (Commutativity)
2. $(\mathbf{u} + \mathbf{v}) + \mathbf{w} = \mathbf{u} + (\mathbf{v} + \mathbf{w})$ (Associativity)
3. There exists a zero vector $\mathbf{0} \in V$ such that $\mathbf{u} + \mathbf{0} = \mathbf{u}$ (Identity Element of Addition)
4. For each $\mathbf{u} \in V$, there exists $-\mathbf{u} \in V$ such that $\mathbf{u} + (-\mathbf{u}) = \mathbf{0}$ (Existence of Inverse Elements)
5. $a \cdot (\mathbf{u} + \mathbf{v}) = a \cdot \mathbf{u} + a \cdot \mathbf{v}$ (Distributivity of Scalar Multiplication with Respect to Vector Addition)
6. $(a + b) \cdot \mathbf{u} = a \cdot \mathbf{u} + b \cdot \mathbf{u}$ (Distributivity of Scalar Addition)
7. $a \cdot (b \cdot \mathbf{u}) = (ab) \cdot \mathbf{u}$ (Associativity of Scalar Multiplication)
8. $1 \cdot \mathbf{u} = \mathbf{u}$ where 1 is the multiplicative identity in $\mathbb{F}$ (Identity Element of Scalar Multiplication)

2   Norms

A norm on a vector space V is a function $\| \cdot \| : V \to \mathbb{R}$ that satisfies the following properties for all $\mathbf{u}, \mathbf{v} \in V$ and $a \in \mathbb{F}$:

1. $\|\mathbf{u}\| \geq 0$, and $\|\mathbf{u}\| = 0$ if and only if $\mathbf{u} = \mathbf{0}$ (Positivity and Definiteness)
2. $\|a\mathbf{u}\| = |a| \cdot \|\mathbf{u}\|$ (Homogeneity)
3. $\|\mathbf{u} + \mathbf{v}\| \leq \|\mathbf{u}\| + \|\mathbf{v}\|$ (Triangle Inequality)

13

A normed vector space is a pair $(V, \|\cdot\|)$. If every Cauchy sequence in V converges to a limit in V, then V is called a Banach space.

3   Convergence

In normed spaces, the notion of convergence is defined through norms. A sequence $\{\mathbf{u}_n\}$ in a normed space V is said to converge to a limit $\mathbf{u} \in V$, if for every $\epsilon > 0$, there exists an integer N such that for all $n \geq N$, the inequality

$$\|\mathbf{u}_n - \mathbf{u}\| < \epsilon$$

is satisfied. This is denoted by $\lim_{n \to \infty} \mathbf{u}_n = \mathbf{u}$.

Convergence is an essential concept in the analysis of infinite-dimensional spaces, where many analytical techniques rely on the properties of sequences and series.

Practice Problems 1

1. Prove that the set of vectors $\{\mathbf{u} + \mathbf{v}, \mathbf{u} - \mathbf{v}, \mathbf{u}\}$ is a vector space over the field $\mathbb{F}$ given that $\mathbf{u}$ and $\mathbf{v}$ are vectors in a vector space V.

2. Show that if $\|\cdot\|$ is a norm on a vector space V, then $\|\mathbf{u} - \mathbf{v}\|$ defines a metric on V.

3. Consider the sequence $\mathbf{u}_n = \frac{1}{n}\mathbf{u}$. Determine if $\mathbf{u}_n$ converges in the normed space V, and find the limit if it exists.

4. Prove that if V is a Banach space, then every absolutely convergent series in V is also convergent.

$$\sum_{1}^{\infty} \|u_n\| < \infty$$

14

Then $\|u_n\| \to 0$ so $\|S_{n+1} - S_n\| = \|u_{n+1}\| \to 0$;   partial sums are a Cauchy sequence.

5. Verify if the function $\|u\| = \max(|u_1|, |u_2|, \ldots, |u_n|)$ is a norm on $\mathbb{R}^n$.

6. Prove that in any normed vector space V, the zero vector is unique.

Answers 1

1. Prove that the set of vectors $\{\mathbf{u} + \mathbf{v}, \mathbf{u} - \mathbf{v}, \mathbf{u}\}$ is a vector space over the field $\mathbb{F}$.

   Solution: To determine if this set forms a vector space, we check for closure under addition and scalar multiplication, and the existence of a zero vector and additive inverses.

   - Closure under addition: For any two vectors in this set, say $(\mathbf{u} + \mathbf{v})$ and $(\mathbf{u} - \mathbf{v})$, their sum is $(\mathbf{u} + \mathbf{v}) + (\mathbf{u} - \mathbf{v}) = 2\mathbf{u}$, which is a linear combination of $\mathbf{u} + \mathbf{v}$, $\mathbf{u} - \mathbf{v}$, $\mathbf{u}$.

   - Closure under scalar multiplication: For any scalar $a \in \mathbb{F}$ and a vector $(\mathbf{u} + \mathbf{v})$, the scalar multiple is $a(\mathbf{u} + \mathbf{v})$, a linear combination of the given vectors.

   - Zero vector: Assume $\mathbf{0} = \mathbf{u} - \mathbf{u}$, showing that $\mathbf{0} \in V$.

   - Additive inverses: For each vector, say $\mathbf{u} + \mathbf{v}$, the inverse is $-(\mathbf{u} + \mathbf{v})$.

   Thus, the set $\{\mathbf{u} + \mathbf{v}, \mathbf{u} - \mathbf{v}, \mathbf{u}\}$ is a vector space under the axioms stated.

2. Show that if $\| \cdot \|$ is a norm on a vector space V, then $\|\mathbf{u} - \mathbf{v}\|$ defines a metric on V.

   Solution: A metric $d(\mathbf{u}, \mathbf{v}) = \|\mathbf{u} - \mathbf{v}\|$ should satisfy non-negativity, identity of indiscernibles, symmetry, and the triangle inequality.

   - Non-negativity: $\|\mathbf{u} - \mathbf{v}\| \geq 0$ for all $\mathbf{u}, \mathbf{v}$.

   - Identity of indiscernibles: $\|\mathbf{u} - \mathbf{v}\| = 0$ iff $\mathbf{u} = \mathbf{v}$.

   - Symmetry: $\|\mathbf{u} - \mathbf{v}\| = \|\mathbf{v} - \mathbf{u}\|$.

   - Triangle Inequality: $\|\mathbf{u} - \mathbf{v}\| \leq \|\mathbf{u} - \mathbf{w}\| + \|\mathbf{w} - \mathbf{v}\|$.

   These properties hold since $\| \cdot \|$ is a norm, confirming $d(\mathbf{u}, \mathbf{v})$ is a metric.

3. Consider the sequence $\mathbf{u}_n = \frac{1}{n}\mathbf{u}$. Determine if $\mathbf{u}_n$ converges in the normed space V, and find the limit if it exists.

   Solution: A sequence $\mathbf{u}_n$ converges to $\mathbf{0} \in V$ if $\lim_{n \to \infty} \|\mathbf{u}_n - \mathbf{0}\| = \lim_{n \to \infty} \|\mathbf{u}_n\| = 0$. Here, $\|\mathbf{u}_n\| = \|\frac{1}{n}\mathbf{u}\| = \frac{1}{n}\|\mathbf{u}\|$. As $n \to \infty$, $\frac{1}{n} \to 0$. Therefore, $\lim_{n \to \infty} \mathbf{u}_n = \mathbf{0}$.

15

4. Prove that if V is a Banach space, then every absolutely convergent series in V is also convergent.

   Solution: Let $\sum_{n=1}^{\infty} \|\mathbf{u}_n\| < \infty$, the series $\sum_{n=1}^{\infty} \mathbf{u}_n$ is absolutely convergent. By the definition of convergence in a normed vector space and completeness of V, the partial sums $\mathbf{S}_n = \sum_{k=1}^{n} \mathbf{u}_k$ form a Cauchy sequence. Completeness implies $\mathbf{S}_n$ converges in V.

5. Verify if the function $\|\mathbf{u}\| = \max(|u_1|, |u_2|, \ldots, |u_n|)$ is a norm on $\mathbb{R}^n$.

   Solution: To qualify as a norm, $\|\mathbf{u}\|$ should satisfy positivity, homogeneity, and the triangle inequality.

   - Positivity: $\|\mathbf{u}\| = \max(|u_i|) \geq 0$ and $\|\mathbf{u}\| = 0$ if $|u_i| = 0$ for all i, i.e., $\mathbf{u} = \mathbf{0}$.

   - Homogeneity: $\|a\mathbf{u}\| = \max(|a||u_i|) = |a| \max(|u_i|) = |a| \|\mathbf{u}\|$.

   - Triangle Inequality: $\|\mathbf{u} + \mathbf{v}\| = \max(|u_i + v_i|) \leq \max(|u_i| + |v_i|) \leq \max(|u_i|) + \max(|v_i|) = \|\mathbf{u}\| + \|\mathbf{v}\|$.

   Thus, it is a norm.

6. Prove that in any normed vector space V, the zero vector is unique.

   Solution: Assume there exist two zero vectors, $\mathbf{0}_1$ and $\mathbf{0}_2$ in V. For any vector $\mathbf{u}$, we have:

   $\mathbf{u} + \mathbf{0}_1 = \mathbf{u}$ and $\mathbf{u} + \mathbf{0}_2 = \mathbf{u}$.

   Consider $\mathbf{0}_1 + \mathbf{0}_2$, since $\mathbf{0}_1$ is zero vector, $\mathbf{0}_1 + \mathbf{0}_2 = \mathbf{0}_2$. Similarly, since $\mathbf{0}_2$ is zero, $\mathbf{0}_1 + \mathbf{0}_2 = \mathbf{0}_1$. Thus $\mathbf{0}_1 = \mathbf{0}_2$, proving uniqueness.

Practice Problems 2

1. Consider a vector space V over $\mathbb{R}$. Show that if $\mathbf{u}, \mathbf{v} \in V$ and $a, b \in \mathbb{R}$, the linear combination $a\mathbf{u} + b\mathbf{v}$ is also in V.

2. Prove that the zero vector in any vector space V is unique.

3. Let $\| \cdot \|$ be a norm on a vector space V. Prove that for any vector $\mathbf{v} \in V$ and any scalar $a \in \mathbb{R}$, the relation $\|a\mathbf{v}\| = |a| \cdot \|\mathbf{v}\|$ holds.

4. Given a normed space $(V, \|\cdot\|)$, demonstrate that the triangle inequality $\|u + v\| \leq \|u\| + \|v\|$ holds for any $u, v \in V$.

5. Consider a sequence $\{u_n\}$ in a normed space that converges to u. Show that the limit is unique.

Suppose $\|u_n - u\| \to 0$ and $\|u_n - v\| \to 0$

Then $\|u - u_n\| \to 0$ so $\|u - v\| \leq \|u - u_n\| + \|u_n - v\|$

$\therefore \|u - v\| = 0 \implies u = v$

6. In a Banach space V, a Cauchy sequence is defined. Prove that every Cauchy sequence converges in V.

Answers 2

1. **Solution:** We have to show that $au + bv \in V$. By the definition of a vector space, V is closed under scalar multiplication and vector addition. Therefore, both au and bv are elements of V since $a, b \in \mathbb{R}$ and $u, v \in V$. Now using vector addition closure, $au + bv \in V$. Hence, the linear combination is in V.

2. **Solution:** Assume two zero vectors 0_1 and 0_2 exist in V. By definition of the zero vector:

$$u + 0_1 = u \quad \text{and} \quad u + 0_2 = u$$

for all $u \in V$. Therefore, we have:

$$0_1 = 0_2 + 0_1 = 0_2$$

Thus, the zero vector is unique.

3. **Solution:** For $\|av\|$, by the properties of a norm:

$$\|av\| = \|a\| \cdot \|v\| = |a| \cdot \|v\|$$

where $\|a\| = |a|$, demonstrating the homogeneity property of norms.

17

4. **Solution:** Consider vectors $\mathbf{u}, \mathbf{v} \in V$. By the definition of a norm:
$$\|\mathbf{u} + \mathbf{v}\| \leq \|\mathbf{u}\| + \|\mathbf{v}\|$$
This results from the norm satisfying the triangle inequality property, which is a fundamental property of norms.

5. **Solution:** Suppose $\{\mathbf{u}_n\}$ converges to $\mathbf{u}$ and $\mathbf{v}$. Then for any $\epsilon > 0$, there exists N such that for all $n \geq N$:
$$\|\mathbf{u}_n - \mathbf{u}\| < \epsilon/2 \quad \text{and} \quad \|\mathbf{u}_n - \mathbf{v}\| < \epsilon/2$$
Applying the triangle inequality:
$$\|\mathbf{u} - \mathbf{v}\| \leq \|\mathbf{u} - \mathbf{u}_n\| + \|\mathbf{u}_n - \mathbf{v}\| < \epsilon/2 + \epsilon/2 = \epsilon$$
Since ϵ is arbitrary, $\|\mathbf{u} - \mathbf{v}\| = 0$, implying $\mathbf{u} = \mathbf{v}$.

6. **Solution:** Let $\{\mathbf{u}_n\}$ be a Cauchy sequence in a Banach space V. For every $\epsilon > 0$, there exists N such that for all $m, n \geq N$:
$$\|\mathbf{u}_n - \mathbf{u}_m\| < \epsilon$$
Since V is complete, $\{\mathbf{u}_n\}$ converges to a limit $\mathbf{u} \in V$.

Practice Problems 3

1. Prove that every vector space V over a field $\mathbb{F}$ has a unique zero vector.

2. Show that the additive inverse of any element in a vector space V is unique.

3. Demonstrate that the norm function $\|\cdot\| : V \to \mathbb{R}$ satisfies the triangle inequality property.

4. Given a normed space $(V, \|\cdot\|)$, show if a sequence $\{\mathbf{u}_n\}$ converges to $\mathbf{u}$, then $\{\|\mathbf{u}_n\|\}$ converges to $\|\mathbf{u}\|$.

5. Prove that scalar multiplication in V is distributive over vector addition.

6. Let $\|\cdot\|$ be a norm on a vector space V. Verify that the norm satisfies positive definiteness.

9. definition

Answers 3

1. **Solution:** Suppose $\mathbf{0}$ and $\mathbf{0}'$ are both zero vectors in V. By the definition of a zero vector, for any $\mathbf{u} \in V$, we have
$$\mathbf{u} + \mathbf{0} = \mathbf{u} \quad \text{and} \quad \mathbf{u} + \mathbf{0}' = \mathbf{u}.$$

So, by substituting $\mathbf{u} = \mathbf{0}'$ in the first equation, we get
$$\mathbf{0}' + \mathbf{0} = \mathbf{0}'.$$

And substituting $\mathbf{u} = \mathbf{0}$ in the second equation, we obtain
$$\mathbf{0} + \mathbf{0}' = \mathbf{0}.$$

Therefore,
$$\mathbf{0} + \mathbf{0}' = \mathbf{0}'$$

which implies
$$\mathbf{0} = \mathbf{0}'.$$

Thus, the zero vector is unique.

2. **Solution:** Let $\mathbf{v} \in V$ and assume $\mathbf{w}$ and $\mathbf{w}'$ are additive inverses of $\mathbf{v}$. Then, by definition,

$$\mathbf{v} + \mathbf{w} = \mathbf{0} \quad \text{and} \quad \mathbf{v} + \mathbf{w}' = \mathbf{0}.$$

Adding $\mathbf{w}'$ to both sides of the first equation yields

$$\mathbf{v} + \mathbf{w} + \mathbf{w}' = \mathbf{0} + \mathbf{w}'.$$

By associativity, we have

$$(\mathbf{v} + \mathbf{w}) + \mathbf{w}' = \mathbf{w}' \to \mathbf{0} + \mathbf{w}' = \mathbf{w}',$$

therefore $\mathbf{w} = \mathbf{w}'$. Hence, the additive inverse is unique.

3. **Solution:** Let $\mathbf{u}, \mathbf{v} \in V$. The triangle inequality property for a norm states that

$$\|\mathbf{u} + \mathbf{v}\| \leq \|\mathbf{u}\| + \|\mathbf{v}\|.$$

Assume there is a violation: $\|\mathbf{u} + \mathbf{v}\| > \|\mathbf{u}\| + \|\mathbf{v}\|$.

Removing these assumptions results in satisfying the property. By abstracting from here, the initial assumption appears invalid. It follows logically (often through explicit comparison) that any norm satisfies

$$\|\mathbf{u} + \mathbf{v}\| \leq \|\mathbf{u}\| + \|\mathbf{v}\|,$$

thus confirming the truth of the inequality.

4. **Solution:** Suppose $\{\mathbf{u}_n\}$ converges to $\mathbf{u}$ in V, which means:

$$\lim_{n \to \infty} \|\mathbf{u}_n - \mathbf{u}\| = 0.$$

By assumption $\forall \epsilon > 0, \exists N$ such that if $n \geq N$ then,

$$\|\mathbf{u}_n - \mathbf{u}\| < \epsilon.$$

By triangle inequality,

$$\big| \|\mathbf{u}_n\| - \|\mathbf{u}\| \big| \leq \|\mathbf{u}_n - \mathbf{u}\|.$$

Therefore,

$$\big| \|\mathbf{u}_n\| - \|\mathbf{u}\| \big| < \epsilon.$$

This implies that $\|\{\mathbf{u}_n\}|$ converges to $\|\mathbf{u}\|$.

5. **Solution:** Let $a, b \in \mathbb{F}$ and $\mathbf{u}, \mathbf{v} \in V$. The distributivity of scalar multiplication over vector addition requires:

$$a \cdot (\mathbf{u} + \mathbf{v}) = (a \cdot \mathbf{u}) + (a \cdot \mathbf{v}).$$

Consider $\mathbf{w} = \mathbf{u} + \mathbf{v}$. Then by vector space definition

$$a(\mathbf{u} + \mathbf{v}) = a \cdot \mathbf{w}.$$

Using linear property rules:

$$a(\mathbf{u} + \mathbf{v}) = a\mathbf{u} + a\mathbf{v}.$$

Therefore, scalar multiplication distributes over vector addition.

6. **Solution:** Let $\mathbf{u} \in V$. The positive definiteness of a norm states that

$$\|\mathbf{u}\| \geq 0 \quad \text{and} \quad \|\mathbf{u}\| = 0 \Leftrightarrow \mathbf{u} = \mathbf{0}.$$

Assuming $\|\mathbf{u}\| < 0$, contradicts its non-negativity. So, equate $\|\mathbf{u}\| \neq 0$ for any non-zero $\mathbf{u}$.

Thus, by verifying these tags through defined its implications, $\|\mathbf{u}\| \geq 0$.

If $\|\mathbf{u}\| = 0$, proving $\mathbf{u} = \mathbf{0}$ follows swiftly via explorations of norm properties and arithmetic contractions.

Chapter 2

Normed Spaces and Banach Spaces

Normed Vector Spaces

A normed vector space is a vector space V over a field $\mathbb{F}$ equipped with a norm function $\|\cdot\|$, which assigns a non-negative real number to each vector in V. The norm must satisfy the following properties for all vectors $\mathbf{u}, \mathbf{v} \in V$ and scalars $a \in \mathbb{F}$:

1. *Positivity:* $\|\mathbf{u}\| \geq 0$ with equality if and only if $\mathbf{u} = \mathbf{0}$,

$$\|\mathbf{u}\| = 0 \Leftrightarrow \mathbf{u} = \mathbf{0}$$

2. *Homogeneity:* $\|a\mathbf{u}\| = |a| \cdot \|\mathbf{u}\|$,

$$\|a\mathbf{u}\| = |a| \cdot \|\mathbf{u}\|$$

3. *Triangle Inequality:* $\|\mathbf{u} + \mathbf{v}\| \leq \|\mathbf{u}\| + \|\mathbf{v}\|$,

$$\|\mathbf{u} + \mathbf{v}\| \leq \|\mathbf{u}\| + \|\mathbf{v}\|$$

Example: Consider a vector space $\mathbb{R}^n$ with the Euclidean norm defined as $\|\mathbf{x}\|_2 = \sqrt{x_1^2 + x_2^2 + \cdots + x_n^2}$, which checks all norm properties.

Completeness and Cauchy Sequences

A sequence $\{\mathbf{u}_n\}$ in a normed space V is a Cauchy sequence if for every $\epsilon > 0$, there exists an integer N such that for all $m, n \geq N$, the inequality $\|\mathbf{u}_m - \mathbf{u}_n\| < \epsilon$ holds.

$$\forall \epsilon > 0, \exists N \in \mathbb{N} \text{ such that } \|\mathbf{u}_m - \mathbf{u}_n\| < \epsilon \text{ for } m, n \geq N$$

A normed space V is complete, and hence a Banach space, if every Cauchy sequence in V converges to a limit in V.

Example: $\mathbb{R}$ with the absolute value as a norm $|x|$ is a Banach space since every Cauchy sequence of real numbers converges to a real number.

Banach Spaces

A Banach space is a normed vector space that is complete. Completeness allows handling infinite processes, essential when dealing with functions, operators, and quantum theory applications.

Example: The space of continuous functions on the interval $[a, b]$, equipped with the supremum norm $\|f\| = \sup_{x \in [a,b]} |f(x)|$, is a Banach space. This notion facilitates working with convergence and continuity in abstract settings.

Applications in Quantum Theory

In quantum theory, states are often represented as elements of a Banach space. For example, consider the Banach space $L^1(\mathbb{R})$, where each function represents a potential quantum state, ensuring the normalization condition for probability distributions.

Algorithmic computation of quantum observables often involves an understanding of operator norms. To illustrate, given a bounded operator $T : V \to W$ between normed spaces, the operator norm is defined as:

$$\|T\| = \sup_{\|\mathbf{x}\|=1} \|T\mathbf{x}\|$$

Understanding these operator properties underlines more intricate operations encountered in quantum mechanics and field theories.

Examples Relevant to Quantum Theory

Consider the Hilbert space $L^2(\mathbb{R})$, a Banach space with the inner product norm:

$$\|f\|_2 = \left(\int_{-\infty}^{\infty} |f(x)|^2 \, dx \right)^{1/2}$$

Quantum states represented in $L^2(\mathbb{R})$ naturally exhibit completeness, crucial for ensuring the mathematical robustness necessary for physical interpretation. These examples capture completeness and the norm's role in quantum mechanics, reinforcing the notion of states and observables within quantum systems.

Practice Problems 1

1. Prove that every closed subset of a complete metric space is complete. Consider a normed space V and discuss its implications for Banach spaces.

2. Given a normed vector space V over $\mathbb{R}$, demonstrate that the set of continuous linear functionals on V forms a Banach space.

3. Show that every finite-dimensional normed vector space over $\mathbb{R}$ or $\mathbb{C}$ is a Banach space.

4. Consider a sequence $\{f_n\}$ in the Banach space $C([a,b])$ with the supremum norm. Prove that if $\{f_n\}$ converges uniformly to a function f, then f is continuous.

5. Verify that the space $L^p([a,b])$ with $1 \le p < \infty$ is a Banach space by showing convergence of Cauchy sequences.

6. Discuss the implications of the Banach fixed-point theorem and its application to proving existence and uniqueness of solutions to differential equations.

Answers 1

1. **Solution:**

   Let M be a closed subset of a complete metric space X. We need to show that M is complete. Consider a Cauchy sequence $\{x_n\}$ in M. Since X is complete, $\{x_n\}$ converges to some limit $x \in X$. Because M is closed, $x \in M$. Thus, every Cauchy sequence in M converges to a limit in M, showing completeness.

2. **Solution:**

   Let $\mathcal{L}(V)$ denote the set of continuous linear functionals on V. For continuity, consider a Cauchy sequence $\{f_n\}$ in $\mathcal{L}(V)$. By definition, $\|f_n - f_m\| \to 0$ implies for each $x \in V$, $(f_n - f_m)(x) \to 0$. Define f as the pointwise limit. Since f_n are linear and continuous, f is linear. Continuity follows from $\|f_n - f\| \to 0$. Hence, $\mathcal{L}(V)$ is complete and a Banach space.

3. **Solution:**

   Consider a finite-dimensional normed vector space V. Since all norms in finite-dimensional spaces are equivalent, V can be isometrically isomorphic to $\mathbb{R}^n$ or $\mathbb{C}^n$, both of which are complete. Hence, V, being isomorphic to a complete space, is complete itself.

4. **Solution:**

   Suppose $\{f_n\}$ converges uniformly to f in $C([a,b])$. For any $\epsilon > 0$, there exists N such that $\|f_n - f\| < \epsilon/3$ for $n \geq N$. Let $n = N$ and $x, y \in [a,b]$. Then:

   $$|f(x) - f(y)| \leq |f(x) - f_N(x)| + |f_N(x) - f_N(y)| + |f_N(y) - f(y)|$$

   $$< \epsilon/3 + |f_N(x) - f_N(y)| + \epsilon/3$$

   By continuity of f_N, $|f_N(x) - f_N(y)| < \epsilon/3$ for small $|x - y|$. Hence, $|f(x) - f(y)| < \epsilon$, proving f is continuous.

5. **Solution:**

   Consider $L^p([a,b])$ with the norm $\|f\|_p = \left(\int_a^b |f(x)|^p dx\right)^{1/p}$. Let $\{f_n\}$ be a Cauchy sequence. For given $\epsilon > 0$, there exists N such that $\|f_n - f_m\|_p < \epsilon$ for $n, m \geq N$. Define f as the pointwise limit. Dominated convergence ensures $\|f_n - f\|_p \to 0$, confirming $f_n \to f$ in $L^p([a,b])$, showing it is complete.

6. **Solution:**

   The Banach fixed-point theorem states if (X, d) is a complete metric space and $T : X \to X$ a contraction, then T has a unique fixed point. Consider differential equations $x'(t) = f(t, x)$ where f is Lipschitz. Define $T(x) = \int_0^t f(s, x(s))ds$. By contractive mappings within a closed interval, fixed points of T correspond to solutions of the differential equation, ensuring existence and uniqueness within X.

Practice Problems 2

1. Consider the vector space $\mathbb{R}^n$ with the Euclidean norm defined as $\|\mathbf{x}\|_2 = \sqrt{x_1^2 + x_2^2 + \cdots + x_n^2}$. Show that this norm satisfies the triangle inequality.

2. Given a normed space V, demonstrate that the zero vector $\mathbf{0}$ is the only vector in V with norm zero.

3. Prove that if $\|\mathbf{u}_n - \mathbf{u}\| \to 0$ as $n \to \infty$, then $\{\mathbf{u}_n\}$ is a Cauchy sequence.

4. Let f be a continuous function on $[a, b]$ and let $\|f\| = \sup_{x \in [a,b]} |f(x)|$. Show that $\|f\|$ defines a norm on the space of continuous functions over $[a, b]$.

5. Verify if the sequence $\{x^n/n!\}$ in the Banach space $L^1(\mathbb{R})$ is a Cauchy sequence.

6. Consider a bounded linear operator $T : V \to W$ between normed spaces. Prove that $\|T\|$, the operator norm, is finite.

Answers 2

1. Consider the vector space $\mathbb{R}^n$ with the Euclidean norm $\|\mathbf{x}\|_2 = \sqrt{x_1^2 + x_2^2 + \cdots + x_n^2}$. Show that this norm satisfies the triangle inequality.

   Solution:

   $$\|\mathbf{u} + \mathbf{v}\|_2 = \sqrt{(u_1 + v_1)^2 + (u_2 + v_2)^2 + \cdots + (u_n + v_n)^2}$$

   By applying the Minkowski inequality, we have:

   $$\leq \sqrt{u_1^2 + u_2^2 + \cdots + u_n^2} + \sqrt{v_1^2 + v_2^2 + \cdots + v_n^2}$$

   This proves that:

   $$\|\mathbf{u} + \mathbf{v}\|_2 \leq \|\mathbf{u}\|_2 + \|\mathbf{v}\|_2$$

2. Given a normed space V, demonstrate that the zero vector $\mathbf{0}$ is the only vector in V with norm zero.

   Solution:

Let $\|\mathbf{v}\| = 0$. Then by the property of positivity:

$$\|\mathbf{v}\| = 0 \Leftrightarrow \mathbf{v} = \mathbf{0}$$

Suppose $\mathbf{v} \neq \mathbf{0}$, then $\|\mathbf{v}\| > 0$. Thus, if $\|\mathbf{v}\| = 0$, it must be that $\mathbf{v} = \mathbf{0}$.

3. Prove that if $\|\mathbf{u}_n - \mathbf{u}\| \to 0$ as $n \to \infty$, then $\{\mathbf{u}_n\}$ is a Cauchy sequence.

   Solution:

   By definition, $\|\mathbf{u}_n - \mathbf{u}\| \to 0$ implies:

   $$\forall \epsilon > 0, \exists N \in \mathbb{N} \text{ such that } \|\mathbf{u}_n - \mathbf{u}\| < \epsilon/2 \text{ for } n \geq N$$

   Consider for $m, n \geq N$:

   $$\|\mathbf{u}_m - \mathbf{u}_n\| \leq \|\mathbf{u}_m - \mathbf{u}\| + \|\mathbf{u} - \mathbf{u}_n\| < \epsilon/2 + \epsilon/2 = \epsilon$$

   Thus, $\{\mathbf{u}_n\}$ is a Cauchy sequence.

4. Let f be a continuous function on $[a, b]$ and let $\|f\| = \sup_{x \in [a,b]} |f(x)|$. Show that $\|f\|$ defines a norm on the space of continuous functions over $[a, b]$.

   Solution:

   Three properties define this as a norm:

   Positivity: $\|f\| \geq 0$ and $\|f\| = 0 \Leftrightarrow f(x) = 0$ for all $x \in [a, b]$.

   Homogeneity: $\|af\| = \sup_{x \in [a,b]} |af(x)| = |a| \sup_{x \in [a,b]} |f(x)| = |a|\|f\|$.

   Triangle Inequality: Let g be another function, $\|f + g\| = \sup_{x \in [a,b]} |f(x) + g(x)| \leq \sup_{x \in [a,b]} |f(x)| + \sup_{x \in [a,b]} |g(x)| = \|f\| + \|g\|$.

5. Verify if the sequence $\{x^n/n!\}$ in the Banach space $L^1(\mathbb{R})$ is a Cauchy sequence.

   Solution:

   Define the sequence: $a_n = x^n/n!$.

   Given $f_n(x) = x^n/n!$, examine the integral:

   $$\int_{-\infty}^{\infty} \left| \frac{x^n}{n!} \right| \, dx \text{ goes to zero as } n \to \infty$$

   Thus, the integral of the absolute difference $|f_n(x) - f_m(x)| \; \forall \epsilon > 0$, verifies the sequence is Cauchy.

6. Consider a bounded linear operator $T : V \to W$ between normed spaces. Prove that $\|T\|$, the operator norm, is finite.

   Solution:

   By definition of $\|T\|$:
   $$\|T\| = \sup_{\|\mathbf{x}\|=1} \|T\mathbf{x}\|$$

   Boundedness implies $\|T\mathbf{x}\| \leq M\|\mathbf{x}\|$ for some $M > 0$.

   Thus, taking the supremum over unit vectors:

   $$\|T\| \leq M$$

   Therefore, $\|T\|$ is finite as boundedness implies M exists.

Practice Problems 3

1. Show that the space of continuous functions on $[a, b]$ with the supremum norm is a Banach space. Verify completeness by considering convergence of Cauchy sequences.

2. Prove that the Euclidean space $\mathbb{R}^n$ with the standard Euclidean norm is a Banach space.

3. For a given bounded linear operator $T : V \to W$ between normed spaces, demonstrate that the operator norm $\|T\|$ satisfies the inequality $\|Tu\| \leq \|T\|\|u\|$ for all $u \in V$.

4. Consider the sequence of functions $\{f_n(x)\} = \{x^n\}$ in the $L^2[0, 1]$ space. Determine if this sequence is a Cauchy sequence and discuss its convergence properties.

5. Verify that for any normed space, $(\mathbb{R}^n, \|\cdot\|)$, the space is a Banach space if and only if it is complete. Use the definition of a Cauchy sequence in a proof.

6. Given a subset U of a Banach space V, what conditions must U satisfy to be a Banach space in its own right? Discuss the necessity of completeness.

Answers 3

1. **Solution:** Consider the space $C([a, b])$ with the supremum norm $\|f\| = \sup_{x \in [a,b]} |f(x)|$.

   (a) A sequence $\{f_n\}$ in $C([a, b])$ is a Cauchy sequence if for every $\epsilon > 0$, there exists N such that for all $m, n \geq N$, $\|f_m - f_n\| = \sup_{x \in [a,b]} |f_m(x) - f_n(x)| < \epsilon$.

   (b) Uniform convergence ensures that $\{f_n\}$ converges to a function $f \in C([a, b])$, which ensures the completeness of $C([a, b])$.

   (c) Hence, $C([a, b])$ with the supremum norm is a Banach space.

2. **Solution:** $\mathbb{R}^n$ with the Euclidean norm is defined as $\|x\|_2 = \sqrt{x_1^2 + x_2^2 + \cdots + x_n^2}$.

   (a) Take a Cauchy sequence $\{x_n\}$ in $\mathbb{R}^n$. For every $\epsilon > 0$, there exists N such that for $m, n \geq N$, we have $\|x_m - x_n\|_2 < \epsilon$.

   (b) Each component $\{x_{k,n}\}$ is a Cauchy sequence in $\mathbb{R}$ and converges in $\mathbb{R}$, hence the whole sequence $\{x_n\}$ converges.

   (c) Thus, $\mathbb{R}^n$ is a Banach space.

3. **Solution:** Define the operator norm $\|T\| = \sup_{\|u\|=1} \|Tu\|$.

   (a) For any $u \neq 0$, write $v = \frac{u}{\|u\|}$, then $\|v\| = 1$.

   (b) So, $\|Tu\| = \|T(\|u\|v)\| = \|u\|\|Tv\| \leq \|u\|\|T\|$.

   (c) Hence, $\|Tu\| \leq \|T\|\|u\|$.

4. **Solution:** Consider $\{f_n(x)\} = \{x^n\}$ in $L^2[0, 1]$.

   (a) Calculate $\|f_m - f_n\|_2 = \left(\int_0^1 |x^m - x^n|^2 \, dx \right)^{1/2}$.

   (b) As $n \to \infty$, $f_n(x) \to 0$ for $x \in [0, 1)$ and $f_n(1) = 1$.

   (c) The sequence does not uniformly converge on $[0, 1]$. Hence, $\{f_n(x)\}$ is not Cauchy in $L^2[0, 1]$.

5. **Solution:** A normed space $(\mathbb{R}^n, \|\cdot\|)$ is complete if every Cauchy sequence $\{x_n\}$ converges.

   (a) Given $\{x_n\}$ is Cauchy, for all $\epsilon > 0$, $\exists N$ such that $\|x_m - x_n\| < \epsilon$ for $m, n \geq N$.

   (b) Convergence to a limit x follows if there exists $x \in \mathbb{R}^n$ such that $\|x_n - x\| \to 0$.

   (c) Conclusively, $\mathbb{R}^n$ is a Banach space if it is complete.

6. **Solution:** A subset U of a Banach space V can be a Banach space if:

   (a) U is closed in V. This ensures limits of Cauchy sequences in U remain in U.

   (b) U inherits the norm from V, maintaining all completeness properties.

   (c) Therefore, completeness (i.e., all Cauchy sequences converge within U) is necessary for U to be a Banach space.

Chapter 3

Hilbert Spaces and Inner Product Spaces

Inner Product Spaces

An inner product space is a vector space V over a field $\mathbb{F}$ (either $\mathbb{R}$ or $\mathbb{C}$) equipped with an inner product $\langle \cdot, \cdot \rangle : V \times V \to \mathbb{F}$. This is a function satisfying for all vectors $\mathbf{u}, \mathbf{v}, \mathbf{w} \in V$ and scalars $a \in \mathbb{F}$:

1. **Conjugate Symmetry:**
$$\langle \mathbf{u}, \mathbf{v} \rangle = \overline{\langle \mathbf{v}, \mathbf{u} \rangle}$$

2. **Linearity in the First Argument:**
$$\langle a\mathbf{u} + \mathbf{v}, \mathbf{w} \rangle = a\langle \mathbf{u}, \mathbf{w} \rangle + \langle \mathbf{v}, \mathbf{w} \rangle$$

3. **Positivity:**
$$\langle \mathbf{v}, \mathbf{v} \rangle \geq 0$$

and $\langle \mathbf{v}, \mathbf{v} \rangle = 0$ if and only if $\mathbf{v} = \mathbf{0}$.

An immediate consequence is the norm induced by the inner product, defined as $\|\mathbf{v}\| = \sqrt{\langle \mathbf{v}, \mathbf{v} \rangle}$.

Properties of Hilbert Spaces

Hilbert spaces are complete inner product spaces. Completeness implies every Cauchy sequence in the space converges to a limit within the space, a crucial property for various applications in mathematical analysis and quantum mechanics.

1 Examples

For example, the space $\mathbb{R}^n$ with the standard dot product, $\langle \mathbf{x}, \mathbf{y} \rangle = x_1 y_1 + x_2 y_2 + \cdots + x_n y_n$, is a Hilbert space. Similarly, the space $L^2([a,b])$, functions f such that $\int_a^b |f(x)|^2 \, dx < \infty$, with the inner product

$$\langle f, g \rangle = \int_a^b f(x)\overline{g(x)} \, dx$$

is also a Hilbert space.

Orthonormal Bases

An orthonormal basis of a Hilbert space H is a basis consisting of vectors that are orthogonal and of unit length. The standard example in $\mathbb{R}^n$ would be the set $\{\mathbf{e}_i\}$ with $\mathbf{e}_i$ being vectors with 1 in the i-th coordinate and 0 elsewhere.

For any $\mathbf{v} \in H$, the expansion in terms of an orthonormal basis $\{\phi_i\}$ is given by:

$$\mathbf{v} = \sum_i \langle \mathbf{v}, \phi_i \rangle \phi_i$$

and

$$\|\mathbf{v}\|^2 = \sum_i |\langle \mathbf{v}, \phi_i \rangle|^2$$

due to the Parseval's identity.

Projection Theorems

1 Orthogonal Projection

Given a closed subspace $M \subseteq H$ and any vector $\mathbf{v} \in H$, there exists a unique vector $\mathbf{u} \in M$ such that

$$\mathbf{v} = \mathbf{u} + \mathbf{u}^\perp$$

where $\mathbf{u}^\perp$ is orthogonal to M. The vector $\mathbf{u}$ is called the orthogonal projection of $\mathbf{v}$ onto M, denoted by $P_M(\mathbf{v})$.

2 Best Approximation Theorem

The closest point in M to $\mathbf{v}$ in terms of the norm is the projection $P_M(\mathbf{v})$. Thus, for all $\mathbf{w} \in M$,

$$\|\mathbf{v} - P_M(\mathbf{v})\| \leq \|\mathbf{v} - \mathbf{w}\|$$

Algorithms for Orthogonalization

The Gram-Schmidt process transforms a basis into an orthonormal basis. Given a linearly independent set $\{\mathbf{v}_1, \mathbf{v}_2, \ldots, \mathbf{v}_n\}$, the orthonormal set $\{\mathbf{q}_1, \mathbf{q}_2, \ldots, \mathbf{q}_n\}$ can be constructed as follows:

Algorithm 1: Gram-Schmidt Process

Input: Vectors $\{\mathbf{v}_1, \mathbf{v}_2, \ldots, \mathbf{v}_n\}$
Output: Orthonormal vectors $\{\mathbf{q}_1, \mathbf{q}_2, \ldots, \mathbf{q}_n\}$
for $k = 1$ *to* n **do**
$\quad \mathbf{u}_k = \mathbf{v}_k - \sum_{j=1}^{k-1} \langle \mathbf{v}_k, \mathbf{q}_j \rangle \mathbf{q}_j$
$\quad \mathbf{q}_k = \frac{\mathbf{u}_k}{\|\mathbf{u}_k\|}$

The resulting set $\{\mathbf{q}_1, \mathbf{q}_2, \ldots, \mathbf{q}_n\}$ forms an orthonormal basis for the span of $\{\mathbf{v}_1, \mathbf{v}_2, \ldots, \mathbf{v}_n\}$.

Practice Problems 1

1. Prove that every finite-dimensional inner product space is a Hilbert space. Explain why completeness is automatically satisfied in this case.

2. Consider an inner product space V with an orthonormal basis $\{e_1, e_2, \ldots, e_n\}$. Show that for any vector $\mathbf{v} \in V$, the coefficients in the expansion $\mathbf{v} = \sum_{i=1}^{n} \langle \mathbf{v}, e_i \rangle e_i$ are unique.

3. Show that the space $L^2([0, 1])$ is a Hilbert space. Include in your answer the verification of completeness.

4. Demonstrate the construction of the projection of a vector $\mathbf{v}$ onto a subspace M of a Hilbert space H using the orthogonal decomposition theorem.

5. Apply the Gram-Schmidt process to the basis $\{\mathbf{v}_1, \mathbf{v}_2\}$ in $\mathbb{R}^2$, where $\mathbf{v}_1 = (2, 3)$ and $\mathbf{v}_2 = (1, 1)$. Find the orthonormal basis.

6. Using an orthonormal basis $\{\mathbf{q}_1, \mathbf{q}_2, \mathbf{q}_3\}$, verify Parseval's identity for the vector $\mathbf{v} = (4, 3, 2)$ in $\mathbb{R}^3$ assuming the basis is the standard orthonormal basis.

Answers 1

1. **Solution:** Every finite-dimensional inner product space V over $\mathbb{F}$ can be viewed as $\mathbb{R}^n$ or $\mathbb{C}^n$, known to be complete. Completeness means any Cauchy sequence converges within V, which holds because all sequences and their limits are finite sums of basis elements, and operations on finite sums preserve limits. Thus, finite-dimensional implies completeness and therefore V is a Hilbert space.

2. **Solution:** Given $\mathbf{v} = \sum_{i=1}^{n} \langle \mathbf{v}, \mathbf{e}_i \rangle \mathbf{e}_i$, assume a different representation $\mathbf{v} = \sum_{i=1}^{n} a_i \mathbf{e}_i$. Then $\langle \mathbf{v}, \mathbf{e}_j \rangle = \langle \sum_{i=1}^{n} a_i \mathbf{e}_i, \mathbf{e}_j \rangle = a_j$ due to orthonormality $\langle \mathbf{e}_i, \mathbf{e}_j \rangle = \delta_{ij}$. Uniqueness follows from all $a_i = \langle \mathbf{v}, \mathbf{e}_i \rangle$.

3. **Solution:** $L^2([0,1])$ consists of equivalence classes of square-integrable functions. For completeness, consider a Cauchy sequence (f_n) in L^2, which means $\int_0^1 |f_n(x) - f_m(x)|^2 \, dx \to 0$ as $n, m \to \infty$. Thus, (f_n) converges to some $f \in L^2$ pointwise or almost everywhere, preserving the L^2-integral property under limits, proving completeness.

4. **Solution:** Given $\mathbf{v} \in H$ and a closed subspace $M \subseteq H$, the orthogonal decomposition theorem states $\mathbf{v} = \mathbf{u} + \mathbf{u}^\perp$ where $\mathbf{u} \in M$ and $\mathbf{u}^\perp \in M^\perp$. The projection $P_M(\mathbf{v}) = \mathbf{u}$ satisfies $\langle \mathbf{v} - \mathbf{u}, \mathbf{w} \rangle = 0$ for any $\mathbf{w} \in M$.

5. **Solution:** Applying Gram-Schmidt,

$$\mathbf{u}_1 = \mathbf{v}_1 = (2,3), \quad \mathbf{q}_1 = \frac{\mathbf{u}_1}{\|\mathbf{u}_1\|} = \left(\frac{2}{\sqrt{13}}, \frac{3}{\sqrt{13}} \right)$$

$$\mathbf{u}_2 = \mathbf{v}_2 - \langle \mathbf{v}_2, \mathbf{q}_1 \rangle \mathbf{q}_1 = (1,1) - \left(\frac{5}{\sqrt{13}} \right) \left(\frac{2}{\sqrt{13}}, \frac{3}{\sqrt{13}} \right)$$

$$= \left(1 - \frac{10}{13}, 1 - \frac{15}{13} \right) = \left(\frac{3}{13}, \frac{-2}{13} \right)$$

$$\mathbf{q}_2 = \frac{\mathbf{u}_2}{\|\mathbf{u}_2\|} = \left(\frac{3}{\sqrt{13}}, \frac{-2}{\sqrt{13}} \right)$$

The orthonormal basis is $\left(\frac{2}{\sqrt{13}}, \frac{3}{\sqrt{13}} \right), \left(\frac{3}{\sqrt{13}}, \frac{-2}{\sqrt{13}} \right)$.

6. **Solution:** With $\mathbf{v} = (4,3,2)$ and $\mathbf{q}_1 = (1,0,0), \mathbf{q}_2 = (0,1,0), \mathbf{q}_3 = (0,0,1)$, Parseval's identity states:

$$\|\mathbf{v}\|^2 = \sum_{i=1}^{3} |\langle \mathbf{v}, \mathbf{q}_i \rangle|^2 = |\langle (4,3,2), (1,0,0) \rangle|^2 + |\langle (4,3,2), (0,1,0) \rangle|^2 + |\langle (4,3,2), (0,0,1) \rangle|^2$$

$$= |4|^2 + |3|^2 + |2|^2 = 4^2 + 3^2 + 2^2 = 16 + 9 + 4 = 29$$

and $\|\mathbf{v}\|^2 = 4^2 + 3^2 + 2^2 = 29$, confirming the identity.

Practice Problems 2

1. Prove that any finite-dimensional inner product space is complete, hence a Hilbert space.

2. Show that in an inner product space, the Cauchy-Schwarz inequality holds:

$$|\langle \mathbf{u}, \mathbf{v} \rangle| \leq \|\mathbf{u}\|\|\mathbf{v}\|$$

for all vectors $\mathbf{u}, \mathbf{v}$.

3. Given an orthonormal basis $\{\phi_1, \phi_2, \ldots, \phi_n\}$ for a Hilbert space H, show that any vector $\mathbf{v} \in H$ can be uniquely written as $\mathbf{v} = \sum_i \langle \mathbf{v}, \phi_i \rangle \phi_i$.

4. Verify that the function $\langle f, g \rangle = \int_a^b f(x)\overline{g(x)}\, dx$ defines an inner product on the space $L^2([a, b])$.

5. Determine whether the set of all continuous functions on $[0, 1]$ with the inner product $\langle f, g \rangle = \int_0^1 f(x)g(x)\, dx$ forms a Hilbert space.

6. Use the Gram-Schmidt process to orthonormalize the set of vectors $\{(1, 1, 1), (1, 0, 0), (0, 1, 0)\}$ in $\mathbb{R}^3$.

Answers 2

1. **To prove that any finite-dimensional inner product space is complete:**

   Solution: Every finite-dimensional inner product space V can be expressed as $\mathbb{R}^n$ or $\mathbb{C}^n$ equipped with an inner product. We need to show that any Cauchy sequence in V has a limit in V.

   - Consider a Cauchy sequence $\{\mathbf{v}_k\}$ in V. Since V is finite-dimensional, this sequence also exists in $\mathbb{R}^n$ or $\mathbb{C}^n$.
   - In $\mathbb{R}^n$ or $\mathbb{C}^n$, every Cauchy sequence converges. Therefore, the sequence $\{\mathbf{v}_k\}$ converges to some limit $\mathbf{v} \in V$.

   Hence, V is complete and thus a Hilbert space.

2. **To show the Cauchy-Schwarz inequality:**

   Solution: Let $\mathbf{u}, \mathbf{v} \in V$.

   - Consider the non-negative function:

   $$f(t) = \langle t\mathbf{u} + \mathbf{v}, t\mathbf{u} + \mathbf{v} \rangle \geq 0$$

   - Expanding $f(t)$:

   $$f(t) = t^2 \|\mathbf{u}\|^2 + 2t\mathrm{Re}(\langle \mathbf{u}, \mathbf{v} \rangle) + \|\mathbf{v}\|^2$$

   - Since $f(t) \geq 0$ for all t, the discriminant must satisfy:

   $$(\mathrm{Re}(\langle \mathbf{u}, \mathbf{v} \rangle))^2 - \|\mathbf{u}\|^2 \|\mathbf{v}\|^2 \leq 0$$

   Therefore,

   $$|\langle \mathbf{u}, \mathbf{v} \rangle| \leq \|\mathbf{u}\| \|\mathbf{v}\|$$

3. **To show that any vector $\mathbf{v} \in H$ can be uniquely written as $\mathbf{v} = \sum_i \langle \mathbf{v}, \phi_i \rangle \phi_i$:**

   Solution:

   (a) Given an orthonormal basis $\{\phi_i\}$, each $\mathbf{w} \in H$ is expressed as:

   $$\mathbf{v} = \sum_i c_i \phi_i$$

   (b) Take the inner product with ϕ_j:

   $$\langle \mathbf{v}, \phi_j \rangle = \left\langle \sum_i c_i \phi_i, \phi_j \right\rangle = \sum_i c_i \langle \phi_i, \phi_j \rangle = c_j$$

   (c) Therefore:

   $$\mathbf{v} = \sum_i \langle \mathbf{v}, \phi_i \rangle \phi_i$$

   Thus, the expression is unique owing to linear independence of the basis.

4. **To verify that the given function defines an inner product:**

   Solution:

   - **Conjugate Symmetry:**

   $$\langle f, g \rangle = \overline{\langle g, f \rangle}$$

34

- **Linearity:**
$$\langle af + bg, h \rangle = a\langle f, h \rangle + b\langle g, h \rangle$$

- **Positivity:**
$$\langle f, f \rangle = \int_a^b |f(x)|^2 \, dx \geq 0$$

and $\langle f, f \rangle = 0$ implies $f = 0$ almost everywhere.

Therefore, $\langle f, g \rangle$ is an inner product on $L^2([a,b])$.

5. **To determine if the set of all continuous functions on $[0,1]$ forms a Hilbert space:**
Solution:

- Check completeness. Consider continuous functions $f_n(x) = x^n$. These are Cauchy in $C([0,1])$ with the uniform norm.
- However, $\lim_{n \to \infty} f_n(x)$ is not continuous as it becomes discontinuous at $x = 0$.

Therefore, $C([0,1])$ is not complete under the given inner product, thus not a Hilbert space.

6. **To use the Gram-Schmidt process to orthonormalize the given set:**
Solution:

(a) Start with $\mathbf{v}_1 = (1,1,1)$. Set $\mathbf{q}_1 = \frac{\mathbf{v}_1}{\|\mathbf{v}_1\|} = \frac{1}{\sqrt{3}}(1,1,1)$.

(b) Project $\mathbf{v}_2 = (1,0,0)$ onto $\mathbf{q}_1$:
$$\mathbf{u}_2 = \mathbf{v}_2 - \langle \mathbf{v}_2, \mathbf{q}_1 \rangle \mathbf{q}_1 = (1,0,0) - \frac{1}{3}(1,1,1)$$
$$= \left(\frac{2}{3}, -\frac{1}{3}, -\frac{1}{3} \right)$$

(c) Normalize $\mathbf{u}_2$:
$$\mathbf{q}_2 = \frac{\mathbf{u}_2}{\|\mathbf{u}_2\|} = \frac{1}{\sqrt{\frac{4}{9} + \frac{1}{9} + \frac{1}{9}}} \left(\frac{2}{3}, -\frac{1}{3}, -\frac{1}{3} \right)$$
$$= \left(\frac{2}{\sqrt{6}}, -\frac{1}{\sqrt{6}}, -\frac{1}{\sqrt{6}} \right)$$

(d) For $\mathbf{v}_3 = (0,1,0)$:
$$\mathbf{u}_3 = \mathbf{v}_3 - \langle \mathbf{v}_3, \mathbf{q}_1 \rangle \mathbf{q}_1 - \langle \mathbf{v}_3, \mathbf{q}_2 \rangle \mathbf{q}_2$$
$$= (0,1,0) - \frac{1}{3}(1,1,1) - 0 = \left(-\frac{1}{3}, \frac{2}{3}, -\frac{1}{3} \right)$$

(e) Normalize $\mathbf{u}_3$:
$$\mathbf{q}_3 = \frac{\mathbf{u}_3}{\|\mathbf{u}_3\|} = \left(-\frac{1}{\sqrt{3}}, \frac{2}{\sqrt{3}}, -\frac{1}{\sqrt{3}} \right)$$

Therefore, the orthonormal set is $\left\{ \frac{1}{\sqrt{3}}(1,1,1), \left(\frac{2}{\sqrt{6}}, -\frac{1}{\sqrt{6}}, -\frac{1}{\sqrt{6}} \right), \left(-\frac{1}{\sqrt{3}}, \frac{2}{\sqrt{3}}, -\frac{1}{\sqrt{3}} \right) \right\}$.

Practice Problems 3

1. Prove that the space $\mathbb{C}^n$ with the standard inner product $\langle \mathbf{x}, \mathbf{y} \rangle = \sum_{k=1}^{n} x_k \overline{y_k}$ is an inner product space.

2. Verify that the space of continuous functions on the interval $[0, 1]$ with the inner product $\langle f, g \rangle = \int_0^1 f(x)\overline{g(x)}\, dx$ is a pre-Hilbert space.

3. Given vectors $\mathbf{u} = (1, 1, 1)$ and $\mathbf{v} = (2, 0, 2)$ in $\mathbb{R}^3$, find the orthogonal projection of $\mathbf{u}$ onto $\mathbf{v}$.

4. Show that in $\mathbb{R}^3$, the vector $\mathbf{u} = (3, 1, -1)$ and the orthogonal projection $\mathbf{v} = (2, -1, 0)$ onto vector $\mathbf{v}$ is $\left(\frac{1}{5}, -\frac{1}{5}, 0\right)$.

5. Use the Gram-Schmidt process to find an orthonormal basis for the span of $\mathbf{v}_1 = (1, 0, 1)$, $\mathbf{v}_2 = (1, 1, 0)$ in $\mathbb{R}^3$.

6. Let H be a Hilbert space with an orthonormal basis $\{\phi_1, \phi_2, ...\}$. Show that for any vector $\mathbf{v} \in H$, we have $\|\mathbf{v}\|^2 = \sum_i |\langle \mathbf{v}, \phi_i \rangle|^2$.

Answers 3

1. Prove that the space $\mathbb{C}^n$ with the standard inner product $\langle \mathbf{x}, \mathbf{y} \rangle = \sum_{k=1}^{n} x_k \overline{y_k}$ is an inner product space. **Solution:** To show that $\mathbb{C}^n$ is an inner product space, we need to verify the three properties of the inner product:

   - **Conjugate Symmetry:**

$$\langle \mathbf{x}, \mathbf{y} \rangle = \sum_{k=1}^{n} x_k \overline{y_k} = \overline{\sum_{k=1}^{n} y_k \overline{x_k}} = \overline{\langle \mathbf{y}, \mathbf{x} \rangle}.$$

   - **Linearity in the First Argument:** For vectors $\mathbf{x}, \mathbf{y}, \mathbf{z} \in \mathbb{C}^n$ and scalar $a \in \mathbb{C}$,

$$\langle a\mathbf{x} + \mathbf{y}, \mathbf{z} \rangle = \sum_{k=1}^{n} (ax_k + y_k)\overline{z_k} = a\sum_{k=1}^{n} x_k \overline{z_k} + \sum_{k=1}^{n} y_k \overline{z_k} = a\langle \mathbf{x}, \mathbf{z} \rangle + \langle \mathbf{y}, \mathbf{z} \rangle.$$

   - **Positivity:**

$$\langle \mathbf{x}, \mathbf{x} \rangle = \sum_{k=1}^{n} x_k \overline{x_k} = \sum_{k=1}^{n} |x_k|^2 \geq 0,$$

   and $\langle \mathbf{x}, \mathbf{x} \rangle = 0$ if and only if $\mathbf{x} = \mathbf{0}$ since each $|x_k|^2 \geq 0$.

   Hence, $\mathbb{C}^n$ is an inner product space.

2. Verify that the space of continuous functions on the interval $[0, 1]$ with the inner product $\langle f, g \rangle = \int_0^1 f(x)\overline{g(x)} \, dx$ is a pre-Hilbert space. **Solution:** To verify the inner product properties for the space of continuous functions:

   - **Conjugate Symmetry:**

$$\langle f, g \rangle = \int_0^1 f(x)\overline{g(x)} \, dx = \overline{\int_0^1 g(x)\overline{f(x)} \, dx} = \overline{\langle g, f \rangle}.$$

   - **Linearity in the First Argument:** For functions f, h, g and scalar a,

$$\langle af + h, g \rangle = \int_0^1 (af(x) + h(x))\overline{g(x)} \, dx = a\int_0^1 f(x)\overline{g(x)} \, dx + \int_0^1 h(x)\overline{g(x)} \, dx = a\langle f, g \rangle + \langle h, g \rangle.$$

   - **Positivity:**

$$\langle f, f \rangle = \int_0^1 |f(x)|^2 \, dx \geq 0.$$

   $\langle f, f \rangle = 0$ if and only if $f(x) = 0$ almost everywhere.

Thus, the space of continuous functions with this inner product is a pre-Hilbert space.

3. Given vectors $\mathbf{u} = (1, 1, 1)$ and $\mathbf{v} = (2, 0, 2)$ in $\mathbb{R}^3$, find the orthogonal projection of $\mathbf{u}$ onto $\mathbf{v}$. **Solution:** The orthogonal projection $\text{proj}_{\mathbf{v}}\mathbf{u}$ is given by:

$$\text{proj}_{\mathbf{v}}\mathbf{u} = \frac{\langle \mathbf{u}, \mathbf{v} \rangle}{\langle \mathbf{v}, \mathbf{v} \rangle}\mathbf{v}.$$

First, compute the inner products:

$$\langle \mathbf{u}, \mathbf{v} \rangle = 1 \cdot 2 + 1 \cdot 0 + 1 \cdot 2 = 4,$$

$$\langle \mathbf{v}, \mathbf{v} \rangle = 2 \cdot 2 + 0 \cdot 0 + 2 \cdot 2 = 8.$$

So,

$$\text{proj}_{\mathbf{v}}\mathbf{u} = \frac{4}{8}(2, 0, 2) = \left(\frac{1}{2} \cdot 2, \frac{1}{2} \cdot 0, \frac{1}{2} \cdot 2 \right) = (1, 0, 1).$$

4. Show that in $\mathbb{R}^3$, the vector $\mathbf{u} = (3, 1, -1)$ and the orthogonal projection $\mathbf{v} = (2, -1, 0)$ onto vector $\mathbf{v}$ is $(\frac{1}{5}, -\frac{1}{5}, 0)$. **Solution:** The orthogonal projection of $\mathbf{u} = (3, 1, -1)$ onto $\mathbf{v} = (2, -1, 0)$ is calculated as:

$$\text{proj}_{\mathbf{v}}\mathbf{u} = \frac{\langle \mathbf{u}, \mathbf{v} \rangle}{\langle \mathbf{v}, \mathbf{v} \rangle}\mathbf{v}.$$

Compute the inner products:

$$\langle \mathbf{u}, \mathbf{v} \rangle = 3 \cdot 2 + 1 \cdot (-1) + (-1) \cdot 0 = 6 - 1 = 5,$$

$$\langle \mathbf{v}, \mathbf{v} \rangle = 2 \cdot 2 + (-1) \cdot (-1) + 0 \cdot 0 = 4 + 1 = 5.$$

Therefore:

$$\text{proj}_{\mathbf{v}}\mathbf{u} = \frac{5}{5}(2, -1, 0) = (2, -1, 0).$$

There appears to be a mistake in the statement that needs correction, as demonstrated, $(2, -1, 0)$ is the correct projection vector rather than $(\frac{1}{5}, -\frac{1}{5}, 0)$.

5. Use the Gram-Schmidt process to find an orthonormal basis for the span of $\mathbf{v}_1 = (1, 0, 1)$, $\mathbf{v}_2 = (1, 1, 0)$ in $\mathbb{R}^3$. **Solution:** The Gram-Schmidt process involves orthogonalizing $\mathbf{v}_2$ with respect to $\mathbf{v}_1$:

$$\mathbf{u}_1 = \mathbf{v}_1 = (1, 0, 1).$$

Compute

$$\mathbf{u}_2 = \mathbf{v}_2 - \text{proj}_{\mathbf{u}_1}\mathbf{v}_2 = \mathbf{v}_2 - \frac{\langle \mathbf{v}_2, \mathbf{u}_1 \rangle}{\langle \mathbf{u}_1, \mathbf{u}_1 \rangle}\mathbf{u}_1.$$

Calculate the inner products:

$$\langle \mathbf{v}_2, \mathbf{u}_1 \rangle = 1 \cdot 1 + 1 \cdot 0 + 0 \cdot 1 = 1,$$

$$\langle \mathbf{u}_1, \mathbf{u}_1 \rangle = 1 \cdot 1 + 0 \cdot 0 + 1 \cdot 1 = 2.$$

Therefore,

$$\mathbf{u}_2 = (1, 1, 0) - \frac{1}{2}(1, 0, 1) = (1, 1, 0) - \left(\frac{1}{2}, 0, \frac{1}{2} \right) = \left(\frac{1}{2}, 1, -\frac{1}{2} \right).$$

Normalize the vectors $\mathbf{u}_1$ and $\mathbf{u}_2$:

$$\mathbf{q}_1 = \frac{\mathbf{u}_1}{\|\mathbf{u}_1\|} = \frac{(1, 0, 1)}{\sqrt{2}} = \left(\frac{1}{\sqrt{2}}, 0, \frac{1}{\sqrt{2}} \right),$$

$$\mathbf{q}_2 = \frac{\mathbf{u}_2}{\|\mathbf{u}_2\|} = \frac{(\frac{1}{2}, 1, -\frac{1}{2})}{\sqrt{\frac{1}{4} + 1 + \frac{1}{4}}} = \frac{(\frac{1}{2}, 1, -\frac{1}{2})}{\sqrt{\frac{3}{2}}} = \left(\frac{1}{\sqrt{6}}, \frac{2}{\sqrt{6}}, -\frac{1}{\sqrt{6}} \right).$$

Thus, the orthonormal basis is:

$$\left\{ \left(\frac{1}{\sqrt{2}}, 0, \frac{1}{\sqrt{2}} \right), \left(\frac{1}{\sqrt{6}}, \frac{2}{\sqrt{6}}, -\frac{1}{\sqrt{6}} \right) \right\}.$$

6. Let H be a Hilbert space with an orthonormal basis $\{\phi_1, \phi_2, ...\}$. Show that for any vector $\mathbf{v} \in H$, we have $\|\mathbf{v}\|^2 = \sum_i |\langle \mathbf{v}, \phi_i \rangle|^2$. **Solution:** For any vector $\mathbf{v} \in H$,

$$\mathbf{v} = \sum_i \langle \mathbf{v}, \phi_i \rangle \phi_i$$

by the property of orthonormal bases. By Parseval's identity in Hilbert spaces,

$$\|\mathbf{v}\|^2 = \langle \mathbf{v}, \mathbf{v} \rangle = \Big\langle \sum_i \langle \mathbf{v}, \phi_i \rangle \phi_i, \sum_j \langle \mathbf{v}, \phi_j \rangle \phi_j \Big\rangle.$$

By orthonormality, $\langle \phi_i, \phi_j \rangle = \delta_{ij}$, we have:

$$\|\mathbf{v}\|^2 = \sum_i |\langle \mathbf{v}, \phi_i \rangle|^2.$$

This completes the demonstration.

Chapter 4

Linear Operators on Hilbert Spaces

Bounded and Unbounded Linear Operators

A linear operator $T : H \to H$ between two Hilbert spaces is called a **bounded** operator if there exists a constant $C \geq 0$ such that:

$$\|T\mathbf{v}\| \leq C\|\mathbf{v}\| \quad \forall \mathbf{v} \in H.$$

The smallest such C is called the **operator norm** of T, denoted $\|T\|$. Thus,

$$\|T\| = \sup_{\|\mathbf{v}\|=1} \|T\mathbf{v}\|$$

An operator is called **unbounded** if it is not bounded. For unbounded operators, their domain $D(T)$ may be a proper subset of H.

1 Examples of Bounded Operators

The identity operator $I : H \to H$, defined by $I\mathbf{v} = \mathbf{v}$, is bounded with $\|I\| = 1$.

Consider the operator $T : \mathbb{R}^n \to \mathbb{R}^n$ defined by $T\mathbf{x} = A\mathbf{x}$ where A is a square matrix. The operator T is bounded if $\|A\|$, the matrix norm induced by the vector norm, is finite.

2 Examples of Unbounded Operators

Operators such as differentiation D, where $Df = f'$, on the space of square-integrable functions are typically unbounded when defined on the full L^2-space. These have domains restricted to differentiable functions.

Adjoint Operators

The **adjoint** of a bounded operator $T : H \to H$ is an operator $T^* : H \to H$ satisfying:

$$\langle T\mathbf{u}, \mathbf{v} \rangle = \langle \mathbf{u}, T^*\mathbf{v} \rangle \quad \forall \mathbf{u}, \mathbf{v} \in H.$$

If T is such that $T^* = T$, T is self-adjoint.

1 Properties of Adjoint Operators

The adjoint operator shares the following properties with T:
  - $\|T^*\| = \|T\|$. - The range of T^* is dense in H if and only if T is densely defined.

Operator Norms and Their Properties

The `operator norm` $\|T\|$ of an operator T provides a measure of the "size" of the operator. For bounded operators, the norm satisfies:
- $\|T\mathbf{v}\| \leq \|T\|\|\mathbf{v}\|$. - $\|ST\| \leq \|S\|\|T\|$ for all bounded operators S, T.

The norm of a self-adjoint operator T is also described in terms of its `spectral radius`:

$$\|T\| = \sup\{|\lambda| : \lambda \in \sigma(T)\}$$

where $\sigma(T)$ denotes the spectrum of T.

Study of Properties in Hilbert Spaces

In a Hilbert space, operators are often studied through their spectra, adjoint, and self-adjoint nature. The `sesquilinear form` introduced by $\langle T\mathbf{u}, \mathbf{v} \rangle$ offers insights into these properties.

1 Spectral Properties

The `spectrum` of an operator T, denoted by $\sigma(T)$, includes all scalars λ such that $T - \lambda I$ is not invertible. The spectral properties are crucial in the spectral theorem.

2 Domain and Range Properties

Operators, especially unbounded ones, have specific domain and range properties that are critical to their behavior. The closure of an operator T is defined, and an operator is called `closed` if it equals its closure.

3 Algorithm for Operator Norm Computation

Algorithm 2: Power Iteration Algorithm for Approximating Operator Norm

---
Input: Operator T, initial vector $\mathbf{v}_0$
Output: Approximation of $\|T\|$
do
  $\quad \mathbf{v}_{k+1} = \frac{T\mathbf{v}_k}{\|T\mathbf{v}_k\|}$;
  $\quad r_k = \langle T\mathbf{v}_{k+1}, \mathbf{v}_{k+1} \rangle$;
while *convergence not achieved*;
Return $\|T\| \approx \sup_k |r_k|$;

By understanding these properties, various significant results in Hilbert space theory can be derived, largely leveraging the operator's relationship with the geometric structure of the space.

Practice Problems 1

1. Prove that the operator norm $\|T\|$ satisfies the inequality $\|T\mathbf{v}\| \leq \|T\|\|\mathbf{v}\|$ for a bounded operator T.

2. Show that the adjoint T^* of any bounded linear operator $T : H \to H$ satisfies $\|T^*\| = \|T\|$.

3. Determine whether the operator defined by $T(\mathbf{v}) = \frac{d\mathbf{v}}{dx}$ acting on the space $L^2([0,1])$ with domain consisting of continuously differentiable functions $C^1([0,1])$ is bounded.

4. Given a Hilbert space H and two bounded linear operators $S, T : H \to H$, prove that $\|ST\| \leq \|S\|\|T\|$.

5. If $T : H \to H$ is a self-adjoint operator, show that its spectral radius $\sup\{|\lambda| : \lambda \in \sigma(T)\}$ is equal to its operator norm $\|T\|$.

6. For a bounded operator T on a Hilbert space H, prove that if the range of T^* is dense in H, the operator T must be densely defined.

Answers 1

1. **Proof:**

   Let $\mathbf{v} \in H$ be such that $\|\mathbf{v}\| \neq 0$. Consider the definition of the operator norm:

   $$\|T\| = \sup_{\|\mathbf{v}\|=1} \|T\mathbf{v}\|$$

   Then, for any $\mathbf{v} \neq 0$, define $\mathbf{w} = \frac{\mathbf{v}}{\|\mathbf{v}\|}$ such that $\|\mathbf{w}\| = 1$. We have

   $$\|T\mathbf{v}\| = \|T(\|\mathbf{v}\|\mathbf{w})\| = \|\mathbf{v}\|\|T\mathbf{w}\| \leq \|\mathbf{v}\|\|T\|$$

   Therefore,

   $$\|T\mathbf{v}\| \leq \|T\|\|\mathbf{v}\|$$

2. **Proof:**

   For a bounded linear operator T, the adjoint T^* satisfies:

   $$\langle T\mathbf{u}, \mathbf{v} \rangle = \langle \mathbf{u}, T^*\mathbf{v} \rangle$$

   Using this relation and the definition of operator norm:

   $$\|T^*\| = \sup_{\|\mathbf{v}\|=1} \|T^*\mathbf{v}\|$$

   $$\|T^*\mathbf{v}\| = \sup_{\|\mathbf{u}\|=1} |\langle T\mathbf{u}, \mathbf{v} \rangle| \leq \|T\|\|\mathbf{v}\|$$

   Thus, $\|T^*\| \leq \|T\|$. Similarly, by considering the adjoint property of T^*:

   $$|\langle T^*\mathbf{v}, \mathbf{u} \rangle| = |\langle \mathbf{v}, T\mathbf{u} \rangle| \leq \|T\|\|\mathbf{u}\|$$

   Therefore, $\|T\| \leq \|T^*\|$. Combining both results, $\|T^*\| = \|T\|$.

3. **Solution:**

   The operator $T(\mathbf{v}) = \frac{d\mathbf{v}}{dx}$ is defined on a domain of continuously differentiable functions $C^1([0,1])$. The norm of $T\mathbf{v}$ can grow arbitrarily large as the derivative $\frac{d\mathbf{v}}{dx}$ increases. The supremum does not exist for all $\mathbf{v} \in L^2([0,1])$ because differentiable functions can be constructed arbitrarily large where the derivative $T\mathbf{v}$ would require an unbounded constant C. Therefore, T is unbounded.

4. **Proof:**

   Consider the definition of operator norms for S and T:

   $$\|ST\mathbf{v}\| \leq \|S\|\|T\mathbf{v}\| \leq \|S\|\|T\|\|\mathbf{v}\|$$

   Therefore, when taking the supremum over unit vectors:

   $$\sup_{\|\mathbf{v}\|=1} \|ST\mathbf{v}\| \leq \|S\|\|T\|$$

   Thus, $\|ST\| \leq \|S\|\|T\|$.

5. **Proof:**

   Assume T is self-adjoint. Then, its spectrum $\sigma(T)$ consists of real eigenvalues. From properties of spectral theory, $\|T\| = \sup_{\|\mathbf{v}\|=1} |\langle T\mathbf{v}, \mathbf{v} \rangle| = \max|\lambda|$, where $\lambda \in \sigma(T)$. Therefore, the spectral radius $\sup\{|\lambda| : \lambda \in \sigma(T)\}$ coincides with $\|T\|$.

6. **Proof:**

   Suppose the range of T^* is dense. Since $T^*\mathbf{u}$ densely spans H, any element in H can be approximated by $T^*\mathbf{u}$. Consequently, T must be densely defined; otherwise, the duality would contradict the density, ensuring that $\mathbf{u}$ exists for all $\mathbf{v}$, proving the relation.

Practice Problems 2

1. Given the operator $T : H \to H$ defined on a Hilbert space H, prove that if T is bounded, then T is continuous.

2. Let $T : H \to H$ be a linear operator with domain $D(T)$. Show that if T is closed and $D(T)$ is dense in H, then T^* is also densely defined.

3. Consider a linear operator T with adjoint T^* on a Hilbert space H. If T is self-adjoint, demonstrate that $\sigma(T) \subset \mathbb{R}$.

4. Demonstrate using an example that the adjoint of an unbounded operator may not be defined on the full space.

5. Provide an example and proof of an unbounded self-adjoint operator and discuss its significance in quantum mechanics.

6. For a given bounded operator T on a Hilbert space H, prove that $\|T\| = \|T^*\|$.

Answers 2

1. Given the operator $T : H \to H$ defined on a Hilbert space H, prove that if T is bounded, then T is continuous.

   Solution: A bounded operator T means there exists a constant $C \geq 0$ such that

   $$\|T\mathbf{v}\| \leq C\|\mathbf{v}\| \quad \forall \mathbf{v} \in H.$$

   To prove continuity, let $\epsilon > 0$. Choose $\delta = \frac{\epsilon}{C}$. Then, for any $\mathbf{u}, \mathbf{v} \in H$ such that $\|\mathbf{u} - \mathbf{v}\| < \delta$, we have:

   $$\|T\mathbf{u} - T\mathbf{v}\| = \|T(\mathbf{u} - \mathbf{v})\| \leq C\|\mathbf{u} - \mathbf{v}\| < C \cdot \frac{\epsilon}{C} = \epsilon.$$

   Thus, T is continuous.

2. Let $T : H \to H$ be a linear operator with domain $D(T)$. Show that if T is closed and $D(T)$ is dense in H, then T^* is also densely defined.

   Solution: We begin by considering the graph $G(T) = \{(\mathbf{v}, T\mathbf{v}) \mid \mathbf{v} \in D(T)\}\}$ which is closed in $H \times H$ due to T being closed. Now, because $D(T)$ is dense in H, for any $\mathbf{w} \in H$, there exists a sequence $\{\mathbf{v}_n\} \subset D(T)$ such that $\mathbf{v}_n \to \mathbf{w}$.

   If $\langle T\mathbf{v}_n, \mathbf{y} \rangle \to \langle \mathbf{u}, \mathbf{y} \rangle$ for some $\mathbf{u} \in H$, then $\mathbf{u}$ defines $T^*\mathbf{y}$. So T^* is densely defined because $\mathbf{u}$ exists for all $\mathbf{y} \in H$.

3. Consider a linear operator T with adjoint T^* on a Hilbert space H. If T is self-adjoint, demonstrate that $\sigma(T) \subset \mathbb{R}$.

   Solution: For a self-adjoint operator T, we have $T^* = T$. Self-adjoint operators have the property $\langle T\mathbf{v}, \mathbf{v} \rangle = \langle \mathbf{v}, T\mathbf{v} \rangle \in \mathbb{R}$ for all $\mathbf{v} \in D(T)$. Now, if $\lambda \in \sigma(T)$, $T - \lambda I$ is not invertible. Suppose $\lambda \notin \mathbb{R}$, then

   $$\langle (T - \lambda I)\mathbf{v}, \mathbf{v} \rangle = \langle T\mathbf{v}, \mathbf{v} \rangle - \lambda \|\mathbf{v}\|^2.$$

   Which does not equal zero for all $\mathbf{v} \neq 0$ since imaginary parts would not cancel, contradicting $\lambda \in \sigma(T)$.

4. Demonstrate using an example that the adjoint of an unbounded operator may not be defined on the full space.

   Solution: Consider $T = -i\frac{d}{dx}$ on $L^2(\mathbb{R})$, defined initially on smooth functions with compact support. The dense domain acts as the core, but the adjoint T^* is defined via integration by parts and is dense but does not cover the full $L^2(\mathbb{R})$ because differentiability is required, limiting domain.

5. Provide an example and proof of an unbounded self-adjoint operator and discuss its significance in quantum mechanics.

   Solution: The position operator X defined by $Xf(x) = xf(x)$ on $L^2(\mathbb{R})$ is self-adjoint with domain $D(X) = \{f \in L^2(\mathbb{R}) \mid xf(x) \in L^2(\mathbb{R})\}$. This operator is pivotal in quantum mechanics as observable quantities, like position or momentum, are related to self-adjoint operators which reflect measurable real values.

6. For a given bounded operator T on a Hilbert space H, prove that $\|T\| = \|T^*\|$.

   Solution: Since T is bounded, use the property:

   $$\|T\| = \sup_{\|\mathbf{v}\|=1} \|T\mathbf{v}\|.$$

   The adjoint T^* satisfies $\langle T\mathbf{v}, \mathbf{w} \rangle = \langle \mathbf{v}, T^*\mathbf{w} \rangle$. Then,

   $$\|T^*\| = \sup_{\|\mathbf{w}\|=1} \|T^*\mathbf{w}\|.$$

   Thus, $\|T\mathbf{v}\| = \|T^*\mathbf{v}\|$ for all $\mathbf{v}$, so $\|T\| = \|T^*\|$.

Practice Problems 3

1. Prove that if an operator $T : H \to H$ on a Hilbert space H is bounded, then its range $\mathrm{Ran}(T)$ is closed if and only if the range of its adjoint T^* is closed.

2. Show that the identity operator $I : H \to H$ is the smallest linear operator in terms of the operator norm. Specifically, prove that for any bounded operator $T : H \to H$, it holds that $\|I\| \leq \|T\|$.

3. Determine if the differentiation operator $Df = f'$ on the space $L^2[0,1]$ is bounded and find the domain $D(D)$ of its definition.

4. Given a self-adjoint operator T on a Hilbert space H, prove that its spectrum $\sigma(T)$ is real.

5. Consider a sequence of bounded operators $\{T_n\}$ converging strongly to a bounded operator T. Prove that if each T_n is self-adjoint, then T is self-adjoint.

6. Calculate the operator norm of the matrix operator $T : \mathbb{R}^2 \to \mathbb{R}^2$ given by $T = \begin{pmatrix} 3 & 0 \\ 0 & 4 \end{pmatrix}$.

Answers 3

1. **Solution:**

    To prove that $\mathrm{Ran}(T)$ is closed if and only if $\mathrm{Ran}(T^*)$ is closed, we use the following argument:

    - If $\mathrm{Ran}(T)$ is closed, the orthogonal complement $\mathrm{Ran}(T)^\perp = \ker(T^*)$ is also closed, implying $\mathrm{Ran}(T^*)$ is closed by the orthogonal decomposition of Hilbert spaces.
    - Conversely, if $\mathrm{Ran}(T^*)$ is closed, $\ker(T^*) = \mathrm{Ran}(T)^\perp$ is closed, and thus $\mathrm{Ran}(T)$ is closed.

    Therefore, $\mathrm{Ran}(T)$ is closed if and only if $\mathrm{Ran}(T^*)$ is closed.

2. **Solution:**

    The identity operator I on H satisfies $\|I\mathbf{v}\| = \|\mathbf{v}\|$ for all $\mathbf{v} \in H$, so $\|I\| = 1$.

    For any bounded operator T, $\|T\mathbf{v}\| \le \|T\|\|\mathbf{v}\|$ implies $\|I\| \le \|T\|$ as it holds for any vector normalized to $\mathbf{v}$.

    Thus, $\|I\| = 1$ is minimal by definition.

3. **Solution:**

    The operator $Df = f'$ is unbounded on $L^2[0,1]$ because differentiation is not a continuous mapping in this space, leading to an unbounded operator norm.

    The domain $D(D)$ typically includes functions in $L^2[0,1]$ that are absolutely continuous and have square-integrable derivatives.

4. **Solution:**

    For a self-adjoint operator T, consider $\langle T\mathbf{v}, \mathbf{v} \rangle$ which is real for all $\mathbf{v} \in H$.

    Given $\lambda \in \sigma(T)$, then $(T - \lambda I)$ is not invertible, implying $\langle (T - \lambda I)\mathbf{v}, \mathbf{v} \rangle = 0$ for some $\mathbf{v}$, leading λ to be a real number.

5. **Solution:**

Suppose $T_n \to T$ and each T_n is self-adjoint. If $T_n\mathbf{u} \to T\mathbf{u}$ and $\langle T_n\mathbf{u}, \mathbf{v}\rangle = \langle \mathbf{u}, T_n\mathbf{v}\rangle$ holds for all $\mathbf{u}, \mathbf{v}$, then:

$$\lim\langle T_n\mathbf{u}, \mathbf{v}\rangle = \langle T\mathbf{u}, \mathbf{v}\rangle = \langle \mathbf{u}, T\mathbf{v}\rangle$$

This implies T remains self-adjoint.

6. **Solution:**

The operator norm for $T = \begin{pmatrix} 3 & 0 \\ 0 & 4 \end{pmatrix}$ is decided by its singular values, which are diagonal elements abs(3) and abs(4).

Therefore, the norm $\|T\| = \max\{3, 4\} = 4$.

Chapter 5

Spectral Theorem for Bounded Operators

Introduction to the Spectral Theorem

The spectral theorem is a cornerstone of functional analysis, particularly important for its application to quantum mechanics. The theorem provides a powerful framework for understanding bounded self-adjoint operators by linking them to spectral measures. Formally, it states that every bounded self-adjoint operator T on a Hilbert space H can be represented in terms of spectral projections.

Bounded Self-Adjoint Operators

An operator $T : H \to H$ is called `self-adjoint` if it equals its adjoint, $T = T^*$. A self-adjoint operator is inherently `bounded` if there exists a constant $C \geq 0$ such that

$$\|T\mathbf{v}\| \leq C\|\mathbf{v}\|, \quad \forall \mathbf{v} \in H.$$

Self-adjoint operators have real spectra $\sigma(T) \subset \mathbb{R}$, crucial for quantum mechanics where observables are represented by such operators.

Spectral Measures and Projection-Valued Measures

Spectral theorem involves spectral measures, which are a special type of measure that assigns orthogonal projections:

$$E : \mathcal{B}(\mathbb{R}) \to \mathrm{Proj}(H),$$

where $\mathcal{B}(\mathbb{R})$ is the Borel σ-algebra on $\mathbb{R}$, and $\mathrm{Proj}(H)$ is the set of orthogonal projections on H. For any bounded self-adjoint operator T, there exists a unique projection-valued measure E_T such that

$$T = \int_{\sigma(T)} \lambda \, dE_T(\lambda),$$

where λ ranges over the spectrum of T.

1   Properties of Spectral Measures

Spectral measures satisfy several essential properties analogous to probability measures, except they take values in projection operators:

1. $E(\emptyset) = 0$. 2. $E(\mathbb{R}) = I$ (identity operator on H). 3. If (A_n) is a sequence of disjoint sets in $\mathcal{B}(\mathbb{R})$, then

$$E\left(\bigcup_{n=1}^{\infty} A_n\right) = \sum_{n=1}^{\infty} E(A_n),$$

with the sum converging in the strong operator topology.

Applications in Quantum Mechanics

In quantum mechanics, observables are associated with self-adjoint operators on a Hilbert space H. The spectral theorem facilitates this association by providing a structure for describing physical quantities:

- Spectral Decomposition: Each observable corresponds to an operator, whose spectral decomposition informs about possible measurement outcomes and their probabilities.

- Probability Measures: If a state of a quantum system is represented by a vector ψ, the spectral measure E_T creates a probability measure μ given by

$$\mu(A) = \langle E_T(A)\psi, \psi \rangle,$$

defining the probability of the system being found in a state corresponding to the spectrum subset A.

Integral Representation of Operators

The integral representation of self-adjoint operators via spectral measures states that for any polynomially bounded function f:

$$f(T) = \int_{\sigma(T)} f(\lambda) \, dE_T(\lambda)$$

This representation facilitates computations involving functions of operators, which are common in quantum mechanics.

1 Functional Calculus

The functional calculus of self-adjoint operators allows one to define operations beyond polynomials. For a bounded Borel function $f : \mathbb{R} \to \mathbb{C}$, the operator $f(T)$ is defined as:

$$f(T) = \int_{\sigma(T)} f(\lambda) \, dE_T(\lambda).$$

This calculus is employed to describe quantum systems evolving with operators representing energy (Hamiltonians).

Key Examples

Consider the multiplication operator M_f on $L^2(\mathbb{R}, \mu)$ defined by:

$$(M_f \psi)(x) = f(x)\psi(x)$$

for some real-valued function f. Here, the spectral measure corresponds to multiplication by characteristic functions of Borel sets. The spectral theorem implies that M_f is self-adjoint, with spectrum range within the values f takes.

1 Example: Position Operator

In $L^2(\mathbb{R})$, the position operator X is defined by $(X\psi)(x) = x\psi(x)$. As a self-adjoint operator, X is representable via the spectral theorem, and its spectral measure, which corresponds simply to the Dirac measure.

Summary of Fundamental Results

- **Spectral Theorem:** Every bounded self-adjoint operator on a Hilbert space admits a spectral representation through a unique projection-valued measure. - **Spectrum:** The spectrum of self-adjoint operators is contained within $\mathbb{R}$. - **Quantum Mechanics Applications:** The theorem underpins much of the quantum mechanical paradigm concerning observables and their measurement probabilities.

Practice Problems 1

1. Prove that the spectrum of a bounded self-adjoint operator T on a Hilbert space H is contained within $\mathbb{R}$.

2. Show that for a bounded self-adjoint operator T, the spectral measure E_T satisfies the property $E_T(A \cap B) = E_T(A)E_T(B)$ for any $A, B \in \mathcal{B}(\mathbb{R})$.

3. Given a spectral measure E and a vector $\psi \in H$, define the probability measure induced by E on a Borel set $A \subset \mathbb{R}$.

4. For a self-adjoint operator T and a scalar function $f : \mathbb{R} \to \mathbb{C}$, prove that the operator $f(T)$ defined via the spectral theorem is also self-adjoint if f is real-valued.

5. Illustrate the spectral theorem with an example of the multiplication operator on $L^2([a,b])$.

6. Explain the role of projection-valued measures in the formulation of the spectral theorem and provide an example demonstrating their use.

Answers 1

1. **Solution:** For a bounded self-adjoint operator T, the spectral theorem assures that the operator can be expressed as

$$T = \int_{\sigma(T)} \lambda \, dE_T(\lambda).$$

Since T is self-adjoint, we have $T = T^*$. By this property, any complex number λ that is an eigenvalue of T must satisfy $\lambda = \overline{\lambda}$, indicating $\lambda \in \mathbb{R}$. Thus, the spectrum $\sigma(T)$ is contained within $\mathbb{R}$.

2. **Solution:** Given the spectral measure $E_T : \mathcal{B}(\mathbb{R}) \to \text{Proj}(H)$, the desired property states:

$$E_T(A \cap B) = E_T(A)E_T(B).$$

From the definition of spectral measures, E_T maps Borel sets to orthogonal projections, meaning $E_T(A)E_T(B) = E_T(B)E_T(A)$. For $A = B$, $E_T(A \cap A) = E_T(A)E_T(A) = E_T(A)^2 = E_T(A)$, confirming idempotence typical of projections. This property reflects orthogonality inherent in E_T.

3. **Solution:** To define a probability measure from the spectral measure E induced by a vector $\psi \in H$, consider:

$$\mu(A) = \langle E(A)\psi, \psi \rangle,$$

where $A \subset \mathcal{B}(\mathbb{R})$. This probability measure assigns the value to each Borel set A, representing the likelihood that measurements fall within the spectrum subset A.

4. **Solution:** For any real-valued function f, we consider $f(T)$ as:

$$f(T) = \int_{\sigma(T)} f(\lambda) \, dE_T(\lambda).$$

Given that f is real-valued, $f(T) = f(T)^*$ must also hold. Since f is real and $dE_T(\lambda)$ is hermitian, $f(T)$ results in a self-adjoint operator. Verify this by noting that for real-valued f, the integral's self-adjoint components maintain the self-adjointness of $f(T)$.

5. **Solution:** Consider the multiplication operator M_f on $L^2([a,b])$:

$$(M_f\psi)(x) = f(x)\psi(x),$$

where f is a continuous function on $[a,b]$. The associated spectral measure E has projections akin to characteristic functions over $\mathcal{B}([a,b])$. Following the spectral theorem, M_f offers a model where $\sigma(M_f)$ corresponds to range(f), illustrating practical spectral decomposition.

6. **Solution:** Projection-valued measures assign projections to Borel sets, enabling operator decomposition into a conjunction of these. They form the mathematical backbone of the spectral theorem. For example, if P_A is a projection for an interval A, $T = \int \lambda \, dE(\lambda)$ elucidates the operator's re-construction by adding up these function projections, thus decomposing T into its spectral components.

Practice Problems 2

1. Given a bounded self-adjoint operator T on a Hilbert space H, prove that the spectrum $\sigma(T)$ is a subset of $\mathbb{R}$.

2. Show that if T is a bounded self-adjoint operator, then its spectral measure E_T satisfies $E_T(A \cap B) = E_T(A)E_T(B)$ for any $A, B \in \mathcal{B}(\mathbb{R})$.

3. Given a bounded self-adjoint operator T, express the operator norm of T in terms of its spectral measure.

4. If ψ is a state vector in a quantum system and T is a bounded self-adjoint operator, show how to compute the expected value $\langle T\psi, \psi \rangle$ using the spectral measure E_T.

5. Demonstrate that the functional calculus for bounded self-adjoint operators provides a consistent way to define e^T for such an operator T.

6. Prove that the position operator X in $L^2(\mathbb{R})$ is self-adjoint, providing a brief explanation of its spectral decomposition.

Answers 2

1. **Solution:** To prove that $\sigma(T) \subset \mathbb{R}$ for bounded self-adjoint operators, consider the following:

$$\text{If } \lambda \notin \mathbb{R}, \text{ then } T - \lambda I \text{ is invertible.}$$

To show invertibility, take $\lambda = \alpha + i\beta$ where $\beta \neq 0$. We need to show that $(T - \lambda I)(T - \overline{\lambda}I)$ is positive definite:

$$= (T - \lambda I)(T - \overline{\lambda}I) = (T - \alpha I)^2 + \beta^2 I.$$

Since $T = T^*$, $(T - \alpha I)$ is self-adjoint. Thus, for any $\mathbf{v} \in H$:

$$\langle (T - \lambda I)(T - \overline{\lambda}I)\mathbf{v}, \mathbf{v} \rangle = \langle (T - \alpha I)^2 \mathbf{v}, \mathbf{v} \rangle + \beta^2 \|\mathbf{v}\|^2 > 0.$$

Consequently, $(T - \lambda I)$ is invertible, which implies that $\lambda \notin \sigma(T)$, confirming that $\sigma(T) \subset \mathbb{R}$.

2. **Solution:** Let E_T be the spectral measure of a bounded self-adjoint operator T. We need to prove $E_T(A \cap B) = E_T(A)E_T(B)$ for disjoint A, B:

Since E is a projection-valued measure, $E(A \cap B)$ corresponds to the intersection of projections,

$$\text{for any disjoint sets } A, B \in \mathcal{B}(\mathbb{R}), \ E(A \cap B) = E(A)E(B).$$

Because projections $E(A)$ and $E(B)$ are orthogonal for disjoint measurable sets A and B, the condition holds by the properties of orthogonal projections.

3. **Solution:** For a bounded self-adjoint operator T, the operator norm $\|T\|$ is:

$$\|T\| = \sup\{|\lambda| : \lambda \in \sigma(T)\}.$$

Via the spectral measure E_T, express T as:

$$T = \int_{\sigma(T)} \lambda \, dE_T(\lambda).$$

The norm:

$$\|T\| = \sup_{\|\mathbf{v}\|=1} \left| \left\langle \int_{\sigma(T)} \lambda \, dE_T(\lambda)\mathbf{v}, \mathbf{v} \right\rangle \right| = \sup_{\|\mathbf{v}\|=1} \int_{\sigma(T)} |\lambda| \, d\mu_{\mathbf{v}}(\lambda),$$

where $d\mu_{\mathbf{v}}(\lambda) = \|E_T(\lambda)\mathbf{v}\|^2$.

4. **Solution:** The expected value $\langle T\psi, \psi \rangle$ is computed using:

$$\langle T\psi, \psi \rangle = \int_{\sigma(T)} \lambda \, d\langle E_T(\lambda)\psi, \psi \rangle.$$

This corresponds to the expectation of the outcome of measuring the observable represented by T, where the state is ψ.

5. **Solution:** The functional calculus for bounded self-adjoint operators allows similar operations to functions of real variables. For e^T where $f(x) = e^x$:

$$e^T = \int_{\sigma(T)} e^\lambda \, dE_T(\lambda),$$

defined via the spectral measure, extending polynomial calculus to more complex functions using the Borel function $f(x) = e^x$.

6. **Solution:** The position operator X in $L^2(\mathbb{R})$ where $(X\psi)(x) = x\psi(x)$:

To show X is self-adjoint, demonstrate that $\langle X\psi_1, \psi_2 \rangle = \langle \psi_1, X\psi_2 \rangle$ for $\psi_1, \psi_2 \in$ domain of X.

Via direct integration, X satisfies self-adjoint conditions. Spectral decomposition of X comes directly from multiplication by characteristic functions of Borel sets, with the spectrum covering $\mathbb{R}$.

Practice Problems 3

1. Define a bounded self-adjoint operator T on a Hilbert space H. Explain the conditions necessary for T to be self-adjoint and provide an example.

2. Describe the concept of a spectral measure and its role in the spectral theorem for bounded operators. How is it connected to the operator T?

3. Show how the spectral theorem allows us to express a bounded self-adjoint operator T as an integral involving its spectral measure. Provide a detailed derivation.

4. Discuss the relevance of spectral decomposition in quantum mechanics. How does it provide information about measurement outcomes?

5. How does functional calculus extend the applications of the spectral theorem to functions of operators? Provide an example involving a polynomially bounded function.

6. Consider the position operator X in quantum mechanics as a self-adjoint operator on $L^2(\mathbb{R})$. Explain how its spectral measure simplifies as the Dirac measure and gives insights into physical measurements.

Answers 3

1. **Solution:** A bounded operator $T : H \to H$ on a Hilbert space H is self-adjoint if $T = T^*$, meaning it equals its adjoint. This requires:

$$\langle T\mathbf{v}, \mathbf{w} \rangle = \langle \mathbf{v}, T\mathbf{w} \rangle \quad \forall \mathbf{v}, \mathbf{w} \in H.$$

An example of a self-adjoint operator is the identity operator I where $I\mathbf{v} = \mathbf{v}$ for all $\mathbf{v} \in H$.

2. **Solution:** A spectral measure E assigns each Borel set $A \subset \mathbb{R}$ an orthogonal projection $E(A)$ on H. It satisfies properties such as $E(\emptyset) = 0$ and $E(\mathbb{R}) = I$. For a bounded self-adjoint operator T, the spectral measure E_T satisfies:

$$T = \int_{\sigma(T)} \lambda \, dE_T(\lambda).$$

This links T to cycles of projections, providing a representation in terms of spectral properties.

3. **Solution:** For bounded self-adjoint T, the spectral theorem states:

$$T = \int_{\sigma(T)} \lambda \, dE_T(\lambda),$$

where λ ranges over T's spectrum $\sigma(T)$. Derivation follows from:

- Define E_T projecting spectral intervals onto H.

- Conceptually express T through E_T's changes capturing λ.
- Integration formalizes T exploiting orthogonality of $E_T(A)$ for various A.

4. **Solution:** Spectral decomposition in quantum mechanics translates an observable to a measurement framework:

$$T = \int \lambda \, dE_T(\lambda)$$

where λ are potential outcomes, and E_T assigns probabilities via $\langle E_T(A)\psi, \psi \rangle$. This represents measurement predictions aligning observed values within physics.

5. **Solution:** Functional calculus extends the spectral theorem, enabling definitions for functions of operators:

$$f(T) = \int_{\sigma(T)} f(\lambda) \, dE_T(\lambda).$$

Example: Given $f(\lambda) = \lambda^2$,

$$f(T) = \int_{\sigma(T)} \lambda^2 \, dE_T(\lambda)$$

provides meaningful operator extensions, useful in energy operator computations.

6. **Solution:** In $L^2(\mathbb{R})$, the position operator X has spectral measure linked to Dirac measures where:

$$\psi(x) = \int x \, dE_X(x)$$

This simplifies computation for position as measurement outcomes collapse pointwise, directly relating to classical position concept explored by spectral theorem insights.

Chapter 6

Self-Adjoint and Unbounded Operators

Unbounded Operators on Hilbert Spaces

An operator $T : \mathcal{D}(T) \subseteq H \to H$ is **unbounded** if its domain $\mathcal{D}(T) \neq H$, potentially leading to undefined behavior outside $\mathcal{D}(T)$. Unbounded operators are common in quantum mechanics, typically representing observables with non-compact spectra. For an operator to be considered densely defined, its domain must be dense in H, which is often assumed for unbounded operators.

Self-Adjointness and Symmetric Operators

A densely defined operator T is **symmetric** if:

$$\langle T\mathbf{v}, \mathbf{w} \rangle = \langle \mathbf{v}, T\mathbf{w} \rangle \quad \forall \mathbf{v}, \mathbf{w} \in \mathcal{D}(T).$$

Symmetry is a precursor to self-adjointness, but a symmetric operator might not extend to a self-adjoint one without additional conditions. Formally, an operator T is **self-adjoint** if:

$$T = T^*,$$

where T^* is the adjoint defined by extending:

$$\langle T\mathbf{v}, \mathbf{w} \rangle = \langle \mathbf{v}, T^*\mathbf{w} \rangle \quad \forall \mathbf{v} \in \mathcal{D}(T), \mathbf{w} \in \mathcal{D}(T^*).$$

Deficiency Indices and Their Role

The notion of **deficiency indices** provides a classification for symmetric operators concerning their self-adjoint extensions. For an operator T, the deficiency spaces are defined as:

$$\mathcal{N}_{\pm} = \{\mathbf{v} \in H \mid T^*\mathbf{v} = \pm i\mathbf{v}\},$$

with deficiency indices $n_{\pm} = \dim \mathcal{N}_{\pm}$. The primary result concerning extensions states:
- A symmetric operator T is self-adjoint if and only if $n_+ = n_- = 0$. - If $n_+ = n_-$, T has self-adjoint extensions.

Spectral Theorem for Unbounded Operators

The `spectral theorem` extends to unbounded self-adjoint operators, playing a critical role in quantum field theory. For a self-adjoint operator T, there exists a spectral measure E_T such that:

$$\langle T\mathbf{v}, \mathbf{w} \rangle = \int_{\sigma(T)} \lambda \, d\langle E_T(\lambda)\mathbf{v}, \mathbf{w} \rangle,$$

where $\sigma(T)$ includes isolated and continuous spectra, crucial in representing physical observables.

Von Neumann's Criteria for Self-Adjointness

Von Neumann characterized self-adjoint operators using deficiency indices. For a symmetric unbounded operator T:
- Self-adjointness is ensured when $n_+ = n_-$.
Consider a densely defined symmetric operator T with adjoint T^*:
$\mathcal{D}(T^*)$ is larger than $\mathcal{D}(T)$, and T^* may not be symmetric. To find self-adjoint extensions, one often examines whether:

$$T \subseteq T^* \subseteq T^{**},$$

concludes $T = T^{**}$ precisely when T is self-adjoint, given that the knowledge about the potential symmetries extends beyond $\sigma(T)$'s boundaries.

Applications in Quantum Field Theory

In quantum field theory, unbounded operators model physical quantities such as position and momentum. Their self-adjoint nature ensures observables have real spectra and meaningful physical expectations. The necessity for unbounded operators reflects the infinite-dimensional dynamics inherent in fields, where spectral theory establishes foundational comprehension:
- Conjugation and bounded operators lead to clear spectral decompositions, containing physical solutions and expectation values relevant to quantum mechanics and its extensions into field theory.
The integration:

$$T = \int_{\sigma(T)} \lambda \, dE_T(\lambda),$$

where E_T resolves unboundedness by mapping complex systems into observable sets approachable by the theoretical methods delineated here. These structures not only validate operator theory but underscore the authentic connections between abstract formalism and empirical insights.

Practice Problems 1

1. Consider an unbounded operator $T : \mathcal{D}(T) \subseteq H \to H$, where $\mathcal{D}(T)$ is dense in a Hilbert space H. Show the conditions under which T is symmetric.

2. Given a symmetric operator T, with deficiency indices n_+ and n_-, establish the criteria for T to be self-adjoint.

3. Derive the spectral theorem for a self-adjoint unbounded operator T in terms of its spectral measure E_T.

4. Demonstrate how Von Neumann's theorem outlines that self-adjointness is characterized by a symmetric operator having equal deficiency indices.

5. Provide a detailed explanation of how unbounded operators are involved in the modeling of observables in quantum mechanics, including their significance.

6. Discuss how the integration $T = \int_{\sigma(T)} \lambda \, dE_T(\lambda)$ aids in the resolution of operator unboundedness and why this is essential in quantum field theory.

Answers 1

1. For an unbounded operator $T : \mathcal{D}(T) \subseteq H \to H$, to be **symmetric**, we must show:

$$\langle T\mathbf{v}, \mathbf{w} \rangle = \langle \mathbf{v}, T\mathbf{w} \rangle \quad \forall \mathbf{v}, \mathbf{w} \in \mathcal{D}(T).$$

This equality ensures that the operation of T respects the Hermitian property of inner products, a requirement for symmetry in a linear operator. Therefore, symmetry is confirmed by showing the above condition holds for all elements within the domain of T.

2. Given a symmetric operator T with deficiency indices n_+ and n_-, T is **self-adjoint** if and only if:

$$n_+ = n_- = 0.$$

When the deficiency indices are both zero, the adjoint operator T^* aligns precisely with T, indicating no further self-adjoint extensions are necessary or possible, thereby making T inherently self-adjoint.

3. The **spectral theorem** for a self-adjoint unbounded operator T is derived by relating the operator to its spectral measure E_T, such that:

$$\langle T\mathbf{v}, \mathbf{w} \rangle = \int_{\sigma(T)} \lambda \, d\langle E_T(\lambda)\mathbf{v}, \mathbf{w} \rangle,$$

where $\sigma(T)$ encompasses both point and continuous spectra, linking T's eigenvalues and value functions through projective measures $E_T(\lambda)$.

4. Von Neumann's **theorem** uses deficiency indices to characterize self-adjoint operators. For a symmetric operator:
   - Self-adjointness arises if $n_+ = n_-$. Therefore, symmetry with equal deficiency indices allows T to be self-adjoint, as extending T is unnecessary when such conditions are met.

5. **Unbounded operators** model quantum observables, like position or momentum, which are central to quantum mechanics and often exhibit non-compact, continuous spectra. They require infinite-dimensional spaces to capture all possible states, rendering them essential for accurately modeling physical quantities where bounded operations fail to account for infinitely-scaled interactions.

6. The integration

$$T = \int_{\sigma(T)} \lambda \, dE_T(\lambda),$$

resolves the unbounded nature of operators by interpreting T as a spectral integral, partitioning $\sigma(T)$ to represent observable eigenvalues. This is crucial in quantum field theory, where complex field interactions necessitate managing operator unboundedness to permit physically meaningful evaluations of quantum states.

Practice Problems 2

1. Consider an unbounded linear operator T on a Hilbert space H. If $\mathcal{D}(T)$ is not the entire space H, what conditions must be satisfied for T to be densely defined?

2. Let T be a symmetric operator on a Hilbert space H. Demonstrate the condition for T to be self-adjoint using its adjoint T^*.

3. For an operator T with deficiency indices n_+ and n_-, explain the role of these indices in determining whether T has self-adjoint extensions.

4. Provide the statement of the spectral theorem for unbounded operators and describe its significance in the context of quantum field theory.

5. Using von Neumann's criteria, characterize when a symmetric operator T on a Hilbert space is self-adjoint.

6. Discuss an example of an unbounded self-adjoint operator representing a physical quantity in quantum field theory, explaining how the spectral theorem applies.

Answers 2

1. Consider an unbounded linear operator T on a Hilbert space H. If $\mathcal{D}(T)$ is not the entire space H, what conditions must be satisfied for T to be densely defined?

   Solution: For an operator T to be densely defined, its domain $\mathcal{D}(T)$ must be dense in H. This means that for every vector $\mathbf{v} \in H$, there exists a sequence $\{\mathbf{v}_n\}$ in $\mathcal{D}(T)$ such that:

   $$\mathbf{v}_n \to \mathbf{v} \text{ as } n \to \infty.$$

   Thus, any vector in H can be approximated arbitrarily closely by vectors in $\mathcal{D}(T)$.

2. Let T be a symmetric operator on a Hilbert space H. Demonstrate the condition for T to be self-adjoint using its adjoint T^*.

   Solution: An operator T is symmetric if for all $\mathbf{v}, \mathbf{w} \in \mathcal{D}(T)$:

   $$\langle T\mathbf{v}, \mathbf{w} \rangle = \langle \mathbf{v}, T\mathbf{w} \rangle.$$

   T is self-adjoint if $T = T^*$, meaning $\mathcal{D}(T) = \mathcal{D}(T^*)$ and:

   $$T\mathbf{v} = T^*\mathbf{v} \text{ for all } \mathbf{v} \in \mathcal{D}(T).$$

3. For an operator T with deficiency indices n_+ and n_-, explain the role of these indices in determining whether T has self-adjoint extensions.

   Solution: The deficiency indices are defined by the dimensions of:

   $$\mathcal{N}_+ = \{\mathbf{v} \in H \mid T^*\mathbf{v} = i\mathbf{v}\}, \quad \mathcal{N}_- = \{\mathbf{v} \in H \mid T^*\mathbf{v} = -i\mathbf{v}\}$$

   with dimensions n_+ and n_-, respectively. T has self-adjoint extensions if and only if $n_+ = n_-$. If $n_+ = n_- = 0$, T itself is self-adjoint.

4. Provide the statement of the spectral theorem for unbounded operators and describe its significance in the context of quantum field theory.

   Solution: The spectral theorem for unbounded self-adjoint operators states that for a self-adjoint operator T, there exists a projection-valued measure E_T such that:

   $$\langle T\mathbf{v}, \mathbf{w} \rangle = \int_{\sigma(T)} \lambda \, d\langle E_T(\lambda)\mathbf{v}, \mathbf{w} \rangle.$$

   In quantum field theory, this theorem is crucial for understanding the spectra of observables as they provide real-valued expectation values for physical quantities.

5. Using von Neumann's criteria, characterize when a symmetric operator T on a Hilbert space is self-adjoint.

   Solution: Von Neumann's criteria state that a symmetric operator T on a Hilbert space H is self-adjoint if and only if the deficiency indices $n_+ = n_-$. Specifically, if $n_+ = n_- = 0$, the operator T itself is self-adjoint. If $n_+ = n_- \neq 0$, T has self-adjoint extensions.

6. Discuss an example of an unbounded self-adjoint operator representing a physical quantity in quantum field theory, explaining how the spectral theorem applies.

   Solution: An example is the momentum operator $P = -i\hbar\frac{d}{dx}$ on $L^2(\mathbb{R})$, the space of square-integrable functions on the real line. The domain is dense and related to the space of differentiable functions that vanish at infinity. P is self-adjoint and represents the momentum of a particle. By the spectral theorem, P has a real spectrum, representing possible momentum values, crucial for the probabilistic interpretation of quantum mechanics.

Practice Problems 3

1. Let T be an unbounded operator on a Hilbert space H. Define what it means for T to be densely defined and explain why this property is significant in quantum mechanics.

2. Prove that if an operator T on a Hilbert space is self-adjoint, then its spectrum is real.

3. Consider a symmetric operator T with deficiency indices $n_+ = n_- = 1$. Discuss the existence of self-adjoint extensions for T and describe under what conditions these extensions exist.

4. For a given symmetric operator T, describe the relationship between T and its adjoint T^*. Specifically, discuss how self-adjoint extensions can be characterized using these relationships.

5. Prove or disprove: If the deficiency indices of a symmetric operator T satisfy $n_+ = n_- = 0$, then T is self-adjoint. Explain each step of your reasoning.

6. Explain how the spectral theorem for unbounded operators aids in the analysis of quantum field theories. Provide an example of its application.

Answers 3

1. **Solution:** An operator $T : \mathcal{D}(T) \subseteq H \to H$ is *densely defined* if its domain $\mathcal{D}(T)$ is dense in H. This means for every vector $\mathbf{f} \in H$, there exists a sequence $\mathbf{v}_n \in \mathcal{D}(T)$ such that $\mathbf{v}_n \to \mathbf{f}$ as $n \to \infty$. This property is crucial in quantum mechanics because it ensures that operators representing observables can approximate any state in H, which is critical for meaningful physical measurements.

2. **Solution:** To prove that the spectrum of a self-adjoint operator T is real, consider any complex number $\lambda \notin \mathbb{R}$. Supposing $\lambda \in \sigma(T)$, then $T - \lambda I$ is not invertible. If $T = T^*$, by the definition of self-adjointness, the resolvent operator $(T - \lambda I)^{-1}$ would exist and map into H. However, for non-real λ, the inner product $\langle (T - \lambda I)\mathbf{v}, \mathbf{v} \rangle \neq 0$ due to the symmetry:

$$\langle T\mathbf{v}, \mathbf{v} \rangle - \lambda \langle \mathbf{v}, \mathbf{v} \rangle = \langle (T - \lambda I)\mathbf{v}, \mathbf{v} \rangle$$

This being non-zero implies $\lambda \notin \sigma(T)$, thus proving the spectrum is real.

3. **Solution:** If T is symmetric and $n_+ = n_- = 1$, then T has self-adjoint extensions. According to Von Neumann's theory, a symmetric operator T can have self-adjoint extensions if and only if $n_+ = n_-$. Since $n_+ = n_- = 1$, there exists an extension of T that is self-adjoint. This extension can be characterized by adding suitable boundary conditions or by constructing operators in larger spaces.

4. **Solution:** For any operator T, the adjoint T^* generally has a larger domain than T. The key property of a symmetric operator T is $T \subseteq T^*$. A symmetric operator T is self-adjoint if $T = T^*$, meaning that it coincides with its adjoint on its domain. Self-adjoint extensions can be found by examining operators S such that $T \subseteq S \subseteq T^*$ and ensuring that $S = S^*$.

5. **Solution:** The statement is true. If a symmetric operator T has deficiency indices $n_+ = n_- = 0$, then the deficiency spaces $\mathcal{N}_+$ and $\mathcal{N}_-$ are both trivial, implying that T is self-adjoint. This follows clearly from Von Neumann's characterization, which asserts that self-adjointness corresponds precisely to having zero deficiency indices.

6. **Solution:** The spectral theorem for unbounded operators is pivotal in quantum field theories as it facilitates the understanding and decomposing of complex operators. Consider the Hamiltonian operator $\hat{H}$ in quantum field theory, which is self-adjoint. The spectral theorem implies that there exists a spectral measure, allowing one to express $\hat{H}$ as an integral over its spectrum. This capability is essential for determining the dynamical evolution of quantum states since it enables analysts to describe the state's evolution in terms of its energy eigenvalues - critical in computing probabilities and expectation values of observables within the field.

Chapter 7

Distributions and Generalized Functions

Introduction to Distributions

Distributions, or generalized functions, extend the concept of classical functions to provide a robust framework for handling discontinuities and singularities, essential in modern analysis and theoretical physics. Unlike standard functions, distributions operate on `test functions`, which are smooth, rapidly decaying functions.

1 Test Functions

Let $\mathcal{D}(\mathbb{R}^n)$ denote the space of `test functions`, consisting of infinitely differentiable functions with compact support on $\mathbb{R}^n$. Consider a function $\varphi \in \mathcal{D}(\mathbb{R}^n)$. These functions have properties ensuring all derivatives vanish at infinity, satisfying:

$$\lim_{\|x\| \to \infty} \varphi(x) = 0.$$

The space $\mathcal{D}(\mathbb{R}^n)$ is essential for defining distributions, as it acts as the canonical backdrop for evaluating generalized functions.

2 Definitions and Basic Properties

A `distribution` T is a continuous linear functional on $\mathcal{D}(\mathbb{R}^n)$. For every $\varphi \in \mathcal{D}(\mathbb{R}^n)$, a distribution assigns a real number $T(\varphi)$:

$$T : \mathcal{D}(\mathbb{R}^n) \to \mathbb{R}.$$

Continuity here is defined with respect to the sequence $\{\varphi_k\} \to \varphi$ in $\mathcal{D}(\mathbb{R}^n)$, ensuring convergence $T(\varphi_k) \to T(\varphi)$.

The Dirac Delta Function

The Dirac delta function, δ, serves as a prototype of distributions. It is defined such that for any test function φ:

$$\delta(\varphi) = \varphi(0).$$

Symbolically, δ approximates:

$$\int_{-\infty}^{\infty} \delta(x)\varphi(x)\,dx = \varphi(0),$$

although δ itself does not conventionally exist as a function. It highlights the singular interaction at a point, crucial in point charge and impulse modeling in physics.

Operations on Distributions

1 Differentiation of Distributions

Distributions allow for differentiation beyond classical differentiability. For a distribution T and a test function φ, the derivative T' is defined by:

$$T'(\varphi) = -T(\varphi').$$

This operation adheres to integration by parts, effectively transferring differentiation from the distribution to the test function, while maintaining linearity.

2 Addition and Scalar Multiplication

The addition of distributions T and S and the scalar multiplication by $a \in \mathbb{R}$ are defined straightforwardly for $\varphi \in \mathcal{D}(\mathbb{R}^n)$:

$$(T + S)(\varphi) = T(\varphi) + S(\varphi),$$

$$(aT)(\varphi) = a \cdot T(\varphi).$$

These operations preserve the linear structure of distributions, vital for extending finite operations to an infinite-dimensional setting.

3 Convolution of Distributions

For distributions T and S, convolution $T * S$ defines a new distribution, assuming one distribution has compact support. The convolution is given by:

$$(T * S)(\varphi) = T(x \mapsto S(y \mapsto \varphi(x + y))).$$

This operation is essential in signal processing and differential equations, where it serves as a mechanism to blur or spread influence across the domain.

Examples and Applications

1 Fundamental Examples

1. **Point Evaluation**: $T_f(\varphi) = \int \varphi(x)f(x)\,dx$ for $f \in L_{\text{loc}}^p(\mathbb{R}^n)$. 2. **The Dirac Delta**: Acts at a singular point for evaluating test functions.

Practice Problems 1

1. Define the action of the Dirac delta function δ on the test function $\varphi(x) = e^{-x^2}$ and compute its value.

$$\delta(\varphi(x) = e^{-x^2})$$

2. Verify the linearity of distributions by considering two test functions $\varphi_1(x) = x^2$ and $\varphi_2(x) = x+1$, and distributions T_1 and T_2 with $T_1(\varphi) = 2\int \varphi(x)\,dx$ and $T_2(\varphi) = 3\int \varphi(x)\,dx$. Calculate $(T_1+T_2)(\varphi_1+\varphi_2)$.

$$(T_1 + T_2)(\varphi_1 + \varphi_2)$$

3. Compute the derivative of the distribution $T(\varphi) = \int_{-\infty}^{\infty} x\varphi(x)\,dx$.

$$T'(\varphi)$$

4. Consider a distribution $S(\varphi) = \int_{-\infty}^{\infty} \sin(x)\varphi(x)\,dx$. Determine the value of $S'(\varphi)$ by differentiating the distribution.

$$S'(\varphi)$$

5. Show that the convolution of the Dirac delta function δ with a test function $\varphi(x)$ produces the test function itself. Verify this for $\varphi(x) = \cos(x)$.

$$(\delta * \varphi)(x)$$

6. Demonstrate the use of distributional derivatives by finding the second derivative of $T(\varphi) = \int_{-\infty}^{\infty} e^{-x^2} \varphi(x) \, dx$.

$$T''(\varphi)$$

Answers 1

1. Define the action of the Dirac delta function δ on the test function $\varphi(x) = e^{-x^2}$ and compute its value.

   Solution: The Dirac delta function is defined such that:

   $$\delta(\varphi) = \varphi(0).$$

   For $\varphi(x) = e^{-x^2}$:

   $$\delta(e^{-x^2}) = e^0 = 1.$$

   Therefore,

   $$\delta(\varphi) = 1.$$

2. Verify the linearity of distributions by considering two test functions $\varphi_1(x) = x^2$ and $\varphi_2(x) = x+1$, and distributions T_1 and T_2 with $T_1(\varphi) = 2 \int \varphi(x) \, dx$ and $T_2(\varphi) = 3 \int \varphi(x) \, dx$. Calculate $(T_1+T_2)(\varphi_1+\varphi_2)$.

   Solution: First calculate each distribution separately:

   $$T_1(\varphi_1 + \varphi_2) = 2 \int (x^2 + x + 1) \, dx = 2 \left(\frac{x^3}{3} + \frac{x^2}{2} + x \right) + C.$$

   $$T_2(\varphi_1 + \varphi_2) = 3 \int (x^2 + x + 1) \, dx = 3 \left(\frac{x^3}{3} + \frac{x^2}{2} + x \right) + C.$$

68

Adding, we have:

$$(T_1 + T_2)(\varphi_1 + \varphi_2) = 2\left(\frac{x^3}{3} + \frac{x^2}{2} + x\right) + 3\left(\frac{x^3}{3} + \frac{x^2}{2} + x\right).$$

Putting together, we simplify:

$$= (2+3)\left(\frac{x^3}{3} + \frac{x^2}{2} + x\right) + C.$$

$$= 5\left(\frac{x^3}{3} + \frac{x^2}{2} + x\right) + C.$$

Therefore,

$$(T_1 + T_2)(\varphi_1 + \varphi_2) \text{ shows linearity by scaling operations and sum.}$$

3. Compute the derivative of the distribution $T(\varphi) = \int_{-\infty}^{\infty} x\varphi(x)\,dx$.

    Solution: The derivative $T'(\varphi)$ of a distribution $T(\varphi) = \int_{-\infty}^{\infty} x\varphi(x)\,dx$ is:

    $$T'(\varphi) = -T(\varphi') = -\int_{-\infty}^{\infty} x\varphi'(x)\,dx.$$

    By integration by parts, let $u = x$ then $du = dx$, and let $dv = \varphi'(x)dx$ then $v = \varphi(x)$:

    $$-[x\varphi(x)]_{-\infty}^{\infty} + \int_{-\infty}^{\infty} \varphi(x)\,dx = -0 + \int_{-\infty}^{\infty} \varphi(x)\,dx.$$

    Therefore,

    $$T'(\varphi) = \int_{-\infty}^{\infty} \varphi(x)\,dx.$$

4. Consider a distribution $S(\varphi) = \int_{-\infty}^{\infty} \sin(x)\varphi(x)\,dx$. Determine the value of $S'(\varphi)$ by differentiating the distribution.

    Solution: The derivative $S'(\varphi)$:

    $$S'(\varphi) = -S(\varphi') = -\int_{-\infty}^{\infty} \sin(x)\varphi'(x)\,dx.$$

    By integration by parts, let $u = \sin(x)$ then $du = \cos(x)dx$, and let $dv = \varphi'(x)dx$ then $v = \varphi(x)$:

    $$-[\sin(x)\varphi(x)]_{-\infty}^{\infty} + \int_{-\infty}^{\infty} \cos(x)\varphi(x)\,dx = \int_{-\infty}^{\infty} \cos(x)\varphi(x)\,dx.$$

    Therefore,

    $$S'(\varphi) = \int_{-\infty}^{\infty} \cos(x)\varphi(x)\,dx.$$

5. Show that the convolution of the Dirac delta function δ with a test function $\varphi(x)$ produces the test function itself. Verify this for $\varphi(x) = \cos(x)$.

    Solution: By definition, the convolution given by:

    $$(\delta * \varphi)(x) = \int_{-\infty}^{\infty} \delta(y)\varphi(x-y)\,dy.$$

    Since $\delta(y) = 0$ everywhere except at $y = 0$, this simplifies to:

    $$\varphi(x-0) = \varphi(x).$$

    For $\varphi(x) = \cos(x)$:

    $$(\delta * \cos(x))(x) = \int_{-\infty}^{\infty} \delta(y)\cos(x-y)\,dy = \cos(x).$$

    Therefore,

    $$(\delta * \varphi)(x) = \varphi(x).$$

6. Demonstrate the use of distributional derivatives by finding the second derivative of $T(\varphi) = \int_{-\infty}^{\infty} e^{-x^2} \varphi(x) \, dx$.

   Solution: The first derivative is:

   $$T'(\varphi) = -\int_{-\infty}^{\infty} e^{-x^2} \varphi'(x) \, dx.$$

   Applying integration by parts with $u = e^{-x^2}$, $dv = \varphi'(x)dx$, which gives no boundary terms due to rapid decay:

   $$\int_{-\infty}^{\infty} -(2x)e^{-x^2} \varphi(x) \, dx.$$

   The second derivative is:

   $$T''(\varphi) = -T'(\varphi') = -\int_{-\infty}^{\infty} (2x)e^{-x^2} \varphi'(x) \, dx.$$

   Further integration by parts with no boundary terms yields:

   $$-\left[(2x)e^{-x^2} \varphi(x) \right] + 2 \int_{-\infty}^{\infty} (e^{-x^2} - 2x^2 e^{-x^2})\varphi(x) \, dx.$$

   Simplifies to:

   $$-\int_{-\infty}^{\infty} e^{-x^2}(2 - 4x^2)\varphi(x) \, dx.$$

   Therefore,

   $$T''(\varphi) = \int_{-\infty}^{\infty} (4x^2 - 2)e^{-x^2} \varphi(x) \, dx.$$

Practice Problems 2

1. Define the following distribution and verify its linearity:

   $$T_f(\varphi) = \int_{\mathbb{R}} \varphi(x) f(x) \, dx$$

   where $f \in L^1_{\text{loc}}(\mathbb{R})$.

2. Show that the Dirac delta function δ is a distribution and find $\delta(\varphi)$ when $\varphi(x) = e^{-x^2}$.

3. Prove that the derivative of a distribution is itself a distribution by using the formula:

$$T'(\varphi) = -T(\varphi')$$

4. Compute the convolution of the Dirac delta function δ with a function $f \in \mathcal{D}(\mathbb{R})$.

5. Verify that the space of test functions $\mathcal{D}(\mathbb{R}^n)$ is dense in $L^p(\mathbb{R}^n)$ for $1 \leq p < \infty$.

6. Show that the multiplication of a distribution by a smooth function with compact support results in another distribution.

Answers 2

1. Define the following distribution and verify its linearity:

$$T_f(\varphi) = \int_{\mathbb{R}} \varphi(x) f(x) \, dx$$

Solution:

Linearity: To verify linearity, take two test functions φ_1, φ_2 and scalars a, b:

$$T_f(a\varphi_1 + b\varphi_2) = \int_{\mathbb{R}} (a\varphi_1(x) + b\varphi_2(x))f(x)\,dx$$

$$= a\int_{\mathbb{R}} \varphi_1(x)f(x)\,dx + b\int_{\mathbb{R}} \varphi_2(x)f(x)\,dx$$

$$= aT_f(\varphi_1) + bT_f(\varphi_2)$$

This confirms that T_f is linear.

2. Show that the Dirac delta function δ is a distribution and find $\delta(\varphi)$ when $\varphi(x) = e^{-x^2}$. **Solution:**

Distribution Property: The Dirac delta function δ acts as:

$$\delta(\varphi) = \varphi(0)$$

This is a linear form. Verify with scalars a, b and functions φ_1, φ_2:

$$\delta(a\varphi_1 + b\varphi_2) = (a\varphi_1 + b\varphi_2)(0) = a\varphi_1(0) + b\varphi_2(0) = a\delta(\varphi_1) + b\delta(\varphi_2)$$

With $\varphi(x) = e^{-x^2}$, evaluate:

$$\delta(e^{-x^2}) = e^{-(0)^2} = 1$$

3. Prove that the derivative of a distribution is itself a distribution using:

$$T'(\varphi) = -T(\varphi')$$

Solution:

Let T be a distribution and φ a test function, then φ' is also in $\mathcal{D}(\mathbb{R})$. By definition:

$$T'(\varphi) = -T(\varphi')$$

This operation is linear and continuous for the convergence in $\mathcal{D}(\mathbb{R})$. Thus, T' is a distribution.

4. Compute the convolution of the Dirac delta function δ with a function $f \in \mathcal{D}(\mathbb{R})$. **Solution:**

Convolution is defined as:

$$(\delta * f)(x) = \int_{\mathbb{R}} \delta(y)f(x - y)\,dy$$

Since $\delta(y)$ satisfies $\delta(\varphi) = \varphi(0)$:

$$(\delta * f)(x) = f(x)$$

Hence, $\delta * f = f$.

5. Verify that the space of test functions $\mathcal{D}(\mathbb{R}^n)$ is dense in $L^p(\mathbb{R}^n)$ for $1 \le p < \infty$. **Solution:**

Approximate an $f \in L^p(\mathbb{R}^n)$ by a sequence $\{\varphi_k\}$ from $\mathcal{D}(\mathbb{R}^n)$ such that $\|\varphi_k - f\|_p \to 0$. Use mollifiers:

$$\eta_\epsilon(x) = \frac{1}{\epsilon^n}\eta\left(\frac{x}{\epsilon}\right)$$

Convolution $f * \eta_\epsilon \to f$ in L^p, where η is smooth and compactly supported. Thus, the density is verified.

6. Show that the multiplication of a distribution by a smooth function with compact support results in another distribution. **Solution:**

Given a distribution T and smooth $\psi \in \mathcal{D}(\mathbb{R})$, define ψT by:

$$(\psi T)(\varphi) = T(\psi\varphi)$$

This is linear:

$$(\psi T)(a\varphi_1 + b\varphi_2) = T(\psi(a\varphi_1 + b\varphi_2)) = aT(\psi\varphi_1) + bT(\psi\varphi_2)$$

Continuity follows from that of T, showing ψT is a distribution.

Practice Problems 3

1. Define the concept of a distribution and differentiate the Dirac delta distribution.

$$T(\varphi) = \delta(\varphi).$$

2. Given a distribution T where $T(\varphi) = \int_{\mathbb{R}} x\varphi(x)\,dx$, determine the first derivative T'.

3. Show that the space of test functions $\mathcal{D}(\mathbb{R}^n)$ is dense in $L^2(\mathbb{R}^n)$.

4. Verify whether the following function is a distribution: $T(\varphi) = \int_{\mathbb{R}} \sin(x)\varphi(x)\,dx$.

5. Prove that the convolution of a distribution T with a test function φ, denoted $(T * \varphi)(x)$, is a smooth function.

6. Consider a sequence of test functions $\{\varphi_n\}$ converging to φ. Demonstrate the continuity of the distribution T such that $T(\varphi_n) \to T(\varphi)$.

Answers 3

1. Define the concept of a distribution and differentiate the Dirac delta distribution.
$$T(\varphi) = \delta(\varphi).$$

   Solution: A distribution T is a continuous linear functional on the space of test functions $\mathcal{D}(\mathbb{R}^n)$. For differentiation, if $T(\varphi) = \delta(\varphi) = \varphi(0)$, the derivative T' is defined by:
$$T'(\varphi) = -T(\varphi') = -\varphi'(0).$$

   Therefore, the derivative of the Dirac delta distribution is defined as:
$$\delta'(\varphi) = -\varphi'(0).$$

2. Given a distribution T where $T(\varphi) = \int_\mathbb{R} x\varphi(x)\,dx$, determine the first derivative T'. **Solution:** Using integration by parts, we have:
$$T'(\varphi) = -T(\varphi') = -\int_\mathbb{R} x\varphi'(x)\,dx.$$

   By integration by parts:
$$\int_\mathbb{R} x\varphi'(x)\,dx = -\int_\mathbb{R} \varphi(x)\,dx,$$

   assuming boundary terms vanish. Thus:
$$T'(\varphi) = \int_\mathbb{R} \varphi(x)\,dx.$$

3. Show that the space of test functions $\mathcal{D}(\mathbb{R}^n)$ is dense in $L^2(\mathbb{R}^n)$. **Solution:** For any $f \in L^2(\mathbb{R}^n)$ and $\epsilon > 0$, there exists a compactly supported smooth function φ such that:
$$\|f - \varphi\|_2 < \epsilon.$$

   This is due to the density of compactly supported functions in L^2, derived from the Stone-Weierstrass theorem. Hence, $\mathcal{D}(\mathbb{R}^n)$ is dense in $L^2(\mathbb{R}^n)$.

4. Verify whether the following function is a distribution: $T(\varphi) = \int_\mathbb{R} \sin(x)\varphi(x)\,dx$. **Solution:** The function T depends linearly on φ and can be integrated with test functions due to their compact support and smoothness. $\sin(x)$ is integrable over compact subsets. Thus, T is a distribution.

5. Prove that the convolution of a distribution T with a test function φ, denoted $(T * \varphi)(x)$, is a smooth function. **Solution:** For $(T * \varphi)(x) = T(y \mapsto \varphi(x - y))$, since test functions are infinitely differentiable, the operation respects smoothness:
$$\partial^\alpha (T * \varphi)(x) = T(y \mapsto \partial_y^\alpha \varphi(x - y)).$$

   Differentiating under the distributional bracket maintains smoothness; thereby, $T * \varphi$ is smooth.

6. Consider a sequence of test functions $\{\varphi_n\}$ converging to φ. Demonstrate the continuity of the distribution T such that $T(\varphi_n) \to T(\varphi)$. **Solution:** By definition, distributions are continuous with respect to the convergence of test functions in $\mathcal{D}(\mathbb{R}^n)$. Thus:

$$\lim_{n \to \infty} T(\varphi_n) = T(\lim_{n \to \infty} \varphi_n) = T(\varphi).$$

This confirms the continuity of T.

Chapter 8

Fourier Analysis in Functional Spaces

Introduction to Fourier Transform in L^2 Spaces

The Fourier transform is a powerful mathematical tool used to translate functions from the time to the frequency domain. For square-integrable functions, the Fourier transform provides a way to decompose the function into its constituent frequencies.

1 Definition of the Fourier Transform

For a function $f \in L^2(\mathbb{R})$, the Fourier transform is defined as:

$$\hat{f}(\xi) = \int_{-\infty}^{\infty} f(x)e^{-2\pi i x \xi}\,dx,$$

where ξ represents the frequency variable. This transform is an integral part of functional analysis due to its role in translating problems into spectral language.

2 Properties of the Fourier Transform

The Fourier transform enjoys several key properties:

1. **Linearity:** For $a, b \in \mathbb{C}$ and $f, g \in L^2(\mathbb{R})$,

$$\widehat{(af + bg)}(\xi) = a\hat{f}(\xi) + b\hat{g}(\xi).$$

2. **Scaling:** If $a \neq 0$,

$$\widehat{f(ax)}(\xi) = \frac{1}{|a|}\hat{f}\left(\frac{\xi}{a}\right).$$

3. **Shift:** For any shift x_0,

$$\widehat{f(x - x_0)}(\xi) = e^{-2\pi i x_0 \xi}\hat{f}(\xi).$$

3 Inversion Formula

The inversion formula for the Fourier transform recovers the original function from its transform, given by:

$$f(x) = \int_{-\infty}^{\infty} \hat{f}(\xi)e^{2\pi i x \xi}\,d\xi.$$

This formula is central in applications, as it assures that the transform is reversible under certain conditions.

Plancherel Theorem

The Plancherel theorem establishes the isometry between L^2 spaces and their corresponding frequency domain representation.

1 Statement of Plancherel Theorem

The theorem states that for any $f \in L^2(\mathbb{R})$, the Fourier transform $\hat{f}$ is also in $L^2(\mathbb{R})$, with the preservation of norms:

$$\int_{-\infty}^{\infty} |f(x)|^2 \, dx = \int_{-\infty}^{\infty} |\hat{f}(\xi)|^2 \, d\xi.$$

This equation implies that the Fourier transform is an L^2-isometry.

2 Applications of Plancherel Theorem

Plancherel's theorem is extensively used in solving partial differential equations (PDEs) by converting convolution operations in physical space into multiplication in the frequency space.

3 Proof of Plancherel Theorem

`Continuity:` Consider a sequence $\{f_n\}$ such that $f_n \to f$ in $L^2(\mathbb{R})$. Then $\hat{f}_n \to \hat{f}$ in $L^2(\mathbb{R})$.
   `Duality:` The duality condition confirms:

$$\langle f, g \rangle = \langle \hat{f}, \hat{g} \rangle,$$

where $\langle \cdot, \cdot \rangle$ denotes the inner product.

Applications to Solving Differential Equations

Solving differential equations frequently involves transforming them into an algebraic form via the Fourier transform for simplification.

1 Heat Equation

Consider the heat equation:

$$\frac{\partial u}{\partial t} = \alpha \frac{\partial^2 u}{\partial x^2}.$$

Taking the Fourier transform in spatial variables, this becomes:

$$\frac{\partial \hat{u}}{\partial t} = -\alpha (2\pi \xi)^2 \hat{u}.$$

This can be solved algebraically and then inverted.

2 Wave Equation

The wave equation:

$$\frac{\partial^2 u}{\partial t^2} = c^2 \frac{\partial^2 u}{\partial x^2},$$

transforms to:

$$\frac{\partial^2 \hat{u}}{\partial t^2} = -(2\pi c\xi)^2 \hat{u},$$

solved through ordinary differential equations in the frequency domain.

Convolution Theorem

The convolution theorem simplifies the analysis of convolution operations, crucial in signal processing and solving integral equations.

1 Convolution and Fourier Transform

For two functions $f, g \in L^2(\mathbb{R})$, the convolution is defined by:

$$(f * g)(x) = \int_{-\infty}^{\infty} f(y)g(x - y)\, dy.$$

The convolution theorem states:

$$\widehat{(f * g)}(\xi) = \hat{f}(\xi)\hat{g}(\xi).$$

This agrees with the algebraic structure in the frequency domain, allowing convolution to become multiplication after transformation.

Practice Problems 1

1. Compute the Fourier transform of the function:

$$f_1(x) = e^{-x^2}.$$

2. Verify the convolution theorem for the functions:

$$f_2(x) = e^{-x^2}, \quad g_2(x) = \delta(x),$$

where $\delta(x)$ is the Dirac delta function.

3. Show that the Fourier transform of a shifted function $f_3(x - x_0)$ is:

$$\widehat{f_3(x - x_0)}(\xi) = e^{-2\pi i x_0 \xi} \hat{f}_3(\xi).$$

4. Solve the ordinary differential equation obtained by applying the Fourier transform to:

$$\frac{d^2 u}{dx^2} = -\alpha^2 u.$$

5. Use the Plancherel theorem to show the equivalence of energy in time and frequency domains for:

$$h(x) = \cos(2\pi x).$$

6. Establish the inversion formula for the Fourier transform and demonstrate it with:

$$f_6(x) = e^{-\pi x^2}.$$

Answers 1

1. Compute the Fourier transform of the function:
$$f_1(x) = e^{-x^2}.$$

   Solution: We use the Gaussian integral property for this computation. The Fourier transform is:
$$\hat{f}_1(\xi) = \int_{-\infty}^{\infty} e^{-x^2} e^{-2\pi i x \xi}\, dx.$$

   Completing the square in the exponent:
$$= \int_{-\infty}^{\infty} e^{-(x^2 + 2\pi i x \xi)}\, dx = e^{-\pi^2 \xi^2} \int_{-\infty}^{\infty} e^{-(x + \pi i \xi)^2}\, dx.$$

   The integral simplifies to a Gaussian integral, giving:
$$\hat{f}_1(\xi) = e^{-\pi^2 \xi^2}.$$

   Therefore, the Fourier transform of e^{-x^2} is $e^{-\pi^2 \xi^2}$.

2. Verify the convolution theorem for the functions:
$$f_2(x) = e^{-x^2}, \quad g_2(x) = \delta(x).$$

   Solution: By the definition of convolution:
$$(f_2 * g_2)(x) = \int_{-\infty}^{\infty} e^{-y^2} \delta(x - y)\, dy = e^{-x^2}.$$

   Taking the Fourier transforms:
$$\widehat{(f_2 * g_2)}(\xi) = \hat{f}_2(\xi) \cdot \hat{g}_2(\xi) = e^{-\pi^2 \xi^2} \cdot 1 = e^{-\pi^2 \xi^2}.$$

   Thus, confirming the convolution theorem holds, as the Fourier transform is consistent.

3. Show that the Fourier transform of a shifted function $f_3(x - x_0)$ is:
$$\widehat{f_3(x - x_0)}(\xi) = e^{-2\pi i x_0 \xi} \hat{f}_3(\xi).$$

   Solution: Applying the shift property:
$$\widehat{f_3(x - x_0)}(\xi) = \int_{-\infty}^{\infty} f_3(x - x_0) e^{-2\pi i x \xi}\, dx.$$

   Substitute $u = x - x_0$:
$$= \int_{-\infty}^{\infty} f_3(u) e^{-2\pi i (u + x_0)\xi}\, du = e^{-2\pi i x_0 \xi} \int_{-\infty}^{\infty} f_3(u) e^{-2\pi i u \xi}\, du.$$

   Hence:
$$= e^{-2\pi i x_0 \xi} \hat{f}_3(\xi).$$

4. Solve the ordinary differential equation obtained by applying the Fourier transform to:
$$\frac{d^2 u}{dx^2} = -\alpha^2 u.$$

   Solution: Taking the Fourier transform:
$$\hat{u}''(\xi) = -\alpha^2 \hat{u}(\xi) \Rightarrow (\hat{u}(\xi))'' + \alpha^2 \hat{u}(\xi) = 0.$$

   Solving the ODE:
$$\hat{u}(\xi) = A\cos(\alpha\xi) + B\sin(\alpha\xi).$$

   The solution is dependent on boundary conditions or $u(x)$.

5. Use the Plancherel theorem to show the equivalence of energy in time and frequency domains for:

$$h(x) = \cos(2\pi x).$$

Solution: Perspective equation:

$$\int_{-\infty}^{\infty} |h(x)|^2 \, dx = \int_{-\infty}^{\infty} |\hat{h}(\xi)|^2 \, d\xi.$$

The Fourier transform of $\cos(2\pi x)$ is:

$$\hat{h}(\xi) = \frac{1}{2}(\delta(\xi - 1) + \delta(\xi + 1)).$$

Thus:

$$\int_{-\infty}^{\infty} |h(x)|^2 \, dx = 1 = \int_{-\infty}^{\infty} \delta(\xi - 1)\delta(\xi + 1) \, d\xi.$$

6. Establish the inversion formula for the Fourier transform and demonstrate it with:

$$f_6(x) = e^{-\pi x^2}.$$

Solution: By the inversion formula:

$$f(x) = \int_{-\infty}^{\infty} \hat{f}(\xi) e^{2\pi i x \xi} \, d\xi.$$

For $f_6(x) = e^{-\pi x^2}$, the Fourier transform $\hat{f}_6(\xi) = e^{-\pi \xi^2}$, hence:

$$f(x) = \int_{-\infty}^{\infty} e^{-\pi \xi^2} e^{2\pi i x \xi} \, d\xi = e^{-\pi x^2}.$$

Confirming the inversion aligns with the original function.

Practice Problems 2

1. Compute the Fourier transform of the Gaussian function:

$$f_1(x) = e^{-\pi x^2}$$

2. Show that the Fourier transform of the derivative $f'(x)$ is given by multiplying the transform of $f(x)$ by $2\pi i \xi$:

$$\widehat{f'}(\xi) = 2\pi i \xi \hat{f}(\xi)$$

3. Verify the scaling property of the Fourier transform for:

$$f_3(x) = e^{-\pi x^2}, \quad a = 2$$

4. Use the convolution theorem to find the Fourier transform of the convolution of two functions:

$$f_4(x) = e^{-\pi x^2}, \quad g(x) = e^{-\pi x^2}$$

5. Solve the heat equation using the Fourier transform:

$$\frac{\partial u}{\partial t} = \alpha \frac{\partial^2 u}{\partial x^2}$$

6. Prove the Plancherel theorem for the Fourier transform on $L^2(\mathbb{R})$:

$$\int_{-\infty}^{\infty} |f(x)|^2 \, dx = \int_{-\infty}^{\infty} |\hat{f}(\xi)|^2 \, d\xi$$

Answers 2

1. Compute the Fourier transform of the Gaussian function:

$$f_1(x) = e^{-\pi x^2}$$

Solution: For a Gaussian, the Fourier transform is known to be another Gaussian:

$$\hat{f}_1(\xi) = \int_{-\infty}^{\infty} e^{-\pi x^2} e^{-2\pi i x \xi} \, dx$$

Completing the square in the exponent and using the identity for the integral of a Gaussian, we find:

$$\hat{f}_1(\xi) = e^{-\pi \xi^2}$$

2. Show that the Fourier transform of the derivative $f'(x)$ is given by:

$$\widehat{f'}(\xi) = 2\pi i \xi \hat{f}(\xi)$$

Solution: Using integration by parts, where $u = e^{-2\pi i x \xi}$ and $dv = f'(x)dx$, we have:

$$\int_{-\infty}^{\infty} f'(x) e^{-2\pi i x \xi} dx = -\int_{-\infty}^{\infty} f(x) \frac{d}{dx}(e^{-2\pi i x \xi}) dx$$

$$= -(-2\pi i \xi) \int_{-\infty}^{\infty} f(x) e^{-2\pi i x \xi} dx$$

$$= 2\pi i \xi \hat{f}(\xi)$$

3. Verify the scaling property for $f_3(x)$:

$$f_3(x) = e^{-\pi x^2}, \quad a = 2$$

Solution: Apply the scaling property $\widehat{f(ax)}(\xi) = \frac{1}{|a|} \hat{f}\left(\frac{\xi}{a}\right)$:

$$\widehat{f_3(2x)}(\xi) = \frac{1}{2} \hat{f}_3\left(\frac{\xi}{2}\right)$$

Given $\hat{f}_3(\xi) = e^{-\pi \xi^2}$, hence,

$$\widehat{f_3(2x)}(\xi) = \frac{1}{2} e^{-\pi\left(\frac{\xi}{2}\right)^2} = \frac{1}{2} e^{-\frac{\pi}{4}\xi^2}$$

4. Use the convolution theorem:

$$f_4(x) = e^{-\pi x^2}, \quad g(x) = e^{-\pi x^2}$$

Solution: The Fourier transform of the convolution $f_4 * g$ is the product of the transforms:

$$\widehat{(f_4 * g)}(\xi) = \hat{f}_4(\xi)\hat{g}(\xi)$$

With $\hat{f}_4(\xi) = e^{-\pi \xi^2}$, we have:

$$\widehat{(f_4 * g)}(\xi) = e^{-\pi \xi^2} \cdot e^{-\pi \xi^2} = e^{-2\pi \xi^2}$$

5. Solve the heat equation:

$$\frac{\partial u}{\partial t} = \alpha \frac{\partial^2 u}{\partial x^2}$$

Solution: Take the Fourier transform in x:

$$\frac{\partial \hat{u}}{\partial t} = -\alpha (2\pi \xi)^2 \hat{u}$$

Solving the ordinary differential equation:

$$\hat{u}(\xi, t) = \hat{u}(\xi, 0) e^{-\alpha(2\pi\xi)^2 t}$$

Inverting the Fourier transform gives:

$$u(x,t) = \int_{-\infty}^{\infty} \hat{u}(\xi, 0) e^{-\alpha(2\pi\xi)^2 t} e^{2\pi i x \xi} \, d\xi$$

6. Prove Plancherel's theorem:

$$\int_{-\infty}^{\infty} |f(x)|^2 \, dx = \int_{-\infty}^{\infty} |\hat{f}(\xi)|^2 \, d\xi$$

Solution: By Parseval's identity, which is a corollary of Plancherel's theorem, and noting the Fourier transform is unitary:

$$\int_{-\infty}^{\infty} f(x)\overline{g(x)} \, dx = \int_{-\infty}^{\infty} \hat{f}(\xi)\overline{\hat{g}(\xi)} \, d\xi$$

Setting $g = f$, reveals:

$$\int_{-\infty}^{\infty} |f(x)|^2 \, dx = \int_{-\infty}^{\infty} |\hat{f}(\xi)|^2 \, d\xi$$

Thus, establishing the L^2-isometry property of the Fourier transform.

Practice Problems 3

1. Show that the Fourier transform of the Gaussian function $f(x) = e^{-\pi x^2}$ is itself.

2. Verify the Plancherel theorem for the function $f(x) = e^{-\pi x^2}$.

3. Solve the heat equation $\frac{\partial u}{\partial t} = \alpha \frac{\partial^2 u}{\partial x^2}$ using the Fourier transform, given the initial condition $u(x, 0) = e^{-\pi x^2}$.

4. Use the convolution theorem to find the convolution $(f * g)(x)$ of the functions $f(x) = e^{-\pi x^2}$ and $g(x) = e^{-\pi x^2}$.

5. For $f(x) = e^{-\pi x^2}$, compute the inversion of its Fourier transform to verify the inversion formula.

6. Evaluate the Fourier transform of the shifted function $f(x - x_0) = e^{-\pi(x-x_0)^2}$.

Answers 3

1. To show that the Fourier transform of $f(x) = e^{-\pi x^2}$ is itself:

$$\hat{f}(\xi) = \int_{-\infty}^{\infty} e^{-\pi x^2} e^{-2\pi i x \xi} \, dx.$$

Completing the square in the exponent gives:

$$e^{-\pi(x^2 + 2ix\xi)} = e^{-\pi((x+i\xi)^2 - \xi^2)} = e^{\pi\xi^2} e^{-\pi(x+i\xi)^2}.$$

This integral evaluates to:

$$\hat{f}(\xi) = e^{\pi\xi^2} \int_{-\infty}^{\infty} e^{-\pi(u^2)} \, du = e^{\pi\xi^2} \cdot 1 = e^{-\pi\xi^2},$$

using the fact that $\int_{-\infty}^{\infty} e^{-\pi u^2} \, du = 1$. Hence, $\hat{f}(\xi) = e^{-\pi\xi^2}$, confirming self-duality.

85

2. Verify Plancherel theorem for $f(x) = e^{-\pi x^2}$:

$$\int_{-\infty}^{\infty} |f(x)|^2 \, dx = \int_{-\infty}^{\infty} e^{-2\pi x^2} \, dx = \frac{1}{\sqrt{2}}.$$

From the previous answer, $\hat{f}(\xi) = e^{-\pi \xi^2}$:

$$\int_{-\infty}^{\infty} |\hat{f}(\xi)|^2 \, d\xi = \int_{-\infty}^{\infty} e^{-2\pi \xi^2} \, d\xi = \frac{1}{\sqrt{2}}.$$

Both sides are equal, confirming the Plancherel theorem.

3. Solve the heat equation for $u(x, 0) = e^{-\pi x^2}$: Take the Fourier transform:

$$\frac{\partial \hat{u}}{\partial t} = -\alpha (2\pi \xi)^2 \hat{u},$$

which solves to:

$$\hat{u}(\xi, t) = \hat{u}(\xi, 0) e^{-4\pi^2 \alpha \xi^2 t}.$$

Knowing $\hat{u}(\xi, 0) = e^{-\pi \xi^2}$, we have:

$$\hat{u}(\xi, t) = e^{-\pi \xi^2} e^{-4\pi^2 \alpha \xi^2 t} = e^{-\pi \xi^2 (1 + 4\pi \alpha t)}.$$

Invert:

$$u(x, t) = \mathcal{F}^{-1}\left(e^{-\pi \xi^2 (1 + 4\pi \alpha t)} \right) = \frac{1}{\sqrt{1 + 4\pi \alpha t}} e^{-\frac{\pi x^2}{1 + 4\pi \alpha t}}.$$

4. Using the convolution theorem for $f(x) = g(x) = e^{-\pi x^2}$:

$$\widehat{(f * g)}(\xi) = \hat{f}(\xi) \hat{g}(\xi) = e^{-\pi \xi^2} e^{-\pi \xi^2} = e^{-2\pi \xi^2}.$$

Invert to find convolution:

$$(f * g)(x) = \mathcal{F}^{-1}(e^{-2\pi \xi^2}) = \frac{1}{\sqrt{2}} e^{-\frac{\pi x^2}{2}},$$

showing the spreading of the Gaussian.

5. Check inversion formula for $f(x) = e^{-\pi x^2}$: We calculated $\hat{f}(\xi)$ and now invert:

$$f(x) = \int_{-\infty}^{\infty} e^{-\pi \xi^2} e^{2\pi i x \xi} \, d\xi.$$

Completing the square,

$$= e^{-\pi x^2} \int_{-\infty}^{\infty} e^{-\pi u^2} \, du / |\mathcal{F}| = e^{-\pi x^2},$$

confirming the inversion formula.

6. Evaluate the Fourier transform of the shifted function:

$$\widehat{f(x - x_0)}(\xi) = \int_{-\infty}^{\infty} e^{-\pi (x - x_0)^2} e^{-2\pi i x \xi} \, dx.$$

Substitute $u = x - x_0$:

$$= e^{-2\pi i x_0 \xi} \int_{-\infty}^{\infty} e^{-\pi u^2} e^{-2\pi i u \xi} \, du,$$

which yields:

$$= e^{-2\pi i x_0 \xi} e^{-\pi \xi^2},$$

showing the result of a phase shift in frequency.

86

Chapter 9

Sobolev Spaces

Introduction to Sobolev Spaces

Sobolev spaces, denoted as $W^{k,p}(\Omega)$, are function spaces that play a fundamental role in the analysis of partial differential equations (PDEs). These spaces encompass functions with derivatives that are integrable to a certain power. The parameter k indicates the order of derivatives considered, while p denotes the Lebesgue space L^p in which these derivatives lie.

Weak Derivatives

In classical calculus, derivatives are defined point-wise. However, within Sobolev spaces, weak derivatives extend this notion to functions that may not be differentiable in the classical sense. A function u is said to possess a weak derivative v if, for any test function $\phi \in C_c^\infty(\Omega)$, the following integration by parts formula holds:

$$\int_\Omega u(x) \frac{\partial \phi}{\partial x_i} \, dx = - \int_\Omega v(x) \phi(x) \, dx.$$

This definition ensures that even discontinuous functions, like the absolute value function, have weak derivatives.

Sobolev Space $W^{1,p}(\Omega)$

The most commonly used Sobolev space is $W^{1,p}(\Omega)$, which includes functions whose first derivatives are in $L^p(\Omega)$. The norm on this space is defined as:

$$\|u\|_{W^{1,p}(\Omega)} = \left(\int_\Omega |u(x)|^p \, dx + \sum_{i=1}^n \int_\Omega \left| \frac{\partial u}{\partial x_i} \right|^p \, dx \right)^{1/p}.$$

For $p = 2$, this becomes a Hilbert space, denoted $H^1(\Omega)$.

Embedding Theorems

Embedding theorems provide conditions under which one Sobolev space is continuously embedded into another space, often a Lebesgue space or a Hölder space. The Sobolev Embedding Theorem states that for certain domains $\Omega \subset \mathbb{R}^n$ and appropriate values of p, the space $W^{k,p}(\Omega)$ is continuously embedded in $L^q(\Omega)$ where

$$q = \frac{np}{n - pk} \quad \text{if } pk < n.$$

For $pk = n$, the embedding is into L^q spaces for all $q < \infty$. These embeddings are crucial for proving the regularity of solutions to PDEs.

Poincaré Inequality

The Poincaré inequality is vital in the analysis of PDEs, providing an estimate of a function in terms of its gradient. For a bounded domain Ω with a sufficiently smooth boundary, and $u \in W_0^{1,p}(\Omega)$, the inequality is expressed as:

$$\|u\|_{L^p(\Omega)} \leq C\|\nabla u\|_{L^p(\Omega)},$$

where C is a constant depending on Ω and p. This result implies that the "size" of u can be controlled by the "size" of its derivative.

Extension Theorems

The extension theorem is instrumental in moving from a bounded domain Ω to the entire space $\mathbb{R}^n$. It states that if Ω has a Lipschitz boundary, any function $u \in W^{k,p}(\Omega)$ can be extended to $\tilde{u} \in W^{k,p}(\mathbb{R}^n)$ such that:

$$\tilde{u} = u \text{ in } \Omega \quad \text{and} \quad \|\tilde{u}\|_{W^{k,p}(\mathbb{R}^n)} \leq C\|u\|_{W^{k,p}(\Omega)},$$

with C as a constant. This allows the application of techniques from Fourier analysis and other global methods.

Sobolev Inequality

The Sobolev inequality is a fundamental result providing bounds on norms of functions in Sobolev spaces. For a function $u \in W^{1,p}(\mathbb{R}^n)$, it holds that:

$$\|u\|_{L^{p^*}(\mathbb{R}^n)} \leq C\|\nabla u\|_{L^p(\mathbb{R}^n)},$$

where $p^* = \frac{np}{n-p}$ is the Sobolev conjugate of p. This inequality is crucial when addressing existence and regularity issues in elliptic equations.

Significance in Field Equations

The significance of Sobolev spaces in field equations lies in their ability to handle weak formulations of PDEs. Using weak derivatives, a PDE can be written in a variational form, facilitating the use of functional analysis tools in proving existence and uniqueness of solutions. Moreover, embedding theorems and inequalities provide the continuity and compactness properties required for these proofs.

Practice Problems 1

1. Prove that any function $u \in W^{1,p}(\Omega)$ has a weak derivative in $L^p(\Omega)$ by verifying the integration by parts formula holds for a piecewise linear function $u(x)$.

2. For $\Omega = (0, 1)$ and $u(x) = x^{1/3}$, determine if $u \in W^{1,2}(\Omega)$ by checking if its weak derivative is in $L^2(\Omega)$.

3. Show that for $u \in W_0^{1,p}(\Omega)$, the Poincaré inequality holds. Provide an example function and verify the inequality.

4. Demonstrate the Sobolev inequality where $n = 3$ and $p = 2$. Calculate the Sobolev conjugate p^* and verify the inequality for the test function $u(x) = e^{-|x|^2}$.

5. Given a bounded domain Ω with a Lipschitz boundary, apply the extension theorem to the function $u(x) = \sin(\pi x)$ defined on $\Omega = (0, 1)$.

6. Use an embedding theorem to show that if $k = 1$ and $p = 2$, then $W^{1,2}(\Omega)$ is continuously embedded into $L^6(\Omega)$ for $\Omega \subset \mathbb{R}^3$.

Answers 1

1. Prove that any function $u \in W^{1,p}(\Omega)$ has a weak derivative in $L^p(\Omega)$ by verifying the integration by parts formula holds for a piecewise linear function $u(x)$.

   Solution: Consider a piecewise linear function u that is continuous except at a finite number of points. For any test function $\phi \in C_c^\infty(\Omega)$:

   $$\int_\Omega u(x) \frac{\partial \phi}{\partial x}\, dx = -\int_\Omega v(x)\phi(x)\, dx.$$

   By choosing discontinuity points as the boundary of subintervals and performing integration by parts on each subinterval, we can conclude that the weak derivative v equals the piecewise derivative of u on each interval.

2. For $\Omega = (0,1)$ and $u(x) = x^{1/3}$, determine if $u \in W^{1,2}(\Omega)$.

   Solution: Calculate the weak derivative:

   $$\text{For } u(x) = x^{1/3}, \text{ classical derivative } u'(x) = \frac{1}{3}x^{-2/3}.$$

   Check if this weak derivative is in $L^2(0,1)$:

   $$\int_0^1 \left(\frac{1}{3}x^{-2/3}\right)^2 dx = \int_0^1 \frac{1}{9}x^{-4/3}\, dx,$$

   which diverges as $x \to 0$. Thus, $u \notin W^{1,2}(0,1)$.

3. Show that for $u \in W_0^{1,p}(\Omega)$, the Poincaré inequality holds. Provide an example function.

   Solution: Let $u(x) = x(1-x)$ in $\Omega = (0,1)$. Then:

   $$u'(x) = 1 - 2x.$$

   Verify:

   $$\|u\|_{L^p(0,1)} = \left(\int_0^1 |x(1-x)|^p\, dx\right)^{1/p},$$

   and

   $$\|\nabla u\|_{L^p(0,1)} = \left(\int_0^1 |1-2x|^p\, dx\right)^{1/p}.$$

   Numerical integration shows $\|u\|_{L^p(0,1)} \le C\|\nabla u\|_{L^p(0,1)}$.

4. Demonstrate the Sobolev inequality where $n = 3$ and $p = 2$.

   Solution: Sobolev conjugate $p^* = \frac{3 \cdot 2}{3-2} = 6$. Verify for $u(x) = e^{-|x|^2}$:

   $$\|u\|_{L^6(\mathbb{R}^3)} = \left(\int_{\mathbb{R}^3} |e^{-|x|^2}|^6\, dx\right)^{1/6}$$

   and

   $$\|\nabla u\|_{L^2(\mathbb{R}^3)} = \left(\int_{\mathbb{R}^3} |\nabla e^{-|x|^2}|^2\, dx\right)^{1/2}.$$

   Analytical calculation confirms the inequality holds.

5. Given a bounded domain Ω, apply the extension theorem to $u(x) = \sin(\pi x)$.

   Solution: The extension theorem allows $u(x) = \sin(\pi x)$ in $\Omega = (0, 1)$ to be extended to $\mathbb{R}$:

$$\tilde{u}(x) = \begin{cases} \sin(\pi x) & \text{for } 0 \leq x \leq 1, \\ 0 & \text{otherwise.} \end{cases}$$

   Check

$$\|\tilde{u}\|_{W^{1,p}(\mathbb{R})} \leq C\|u\|_{W^{1,p}(0,1)}.$$

6. Use an embedding theorem to show that $W^{1,2}(\Omega)$ is continuously embedded into $L^6(\Omega)$.

   Solution: For $\Omega \subset \mathbb{R}^3$, and $W^{1,2}(\Omega) \hookrightarrow L^6(\Omega)$, the embedding condition holds:

$$\|u\|_{L^6(\Omega)} \leq C\|u\|_{W^{1,2}(\Omega)}.$$

   Take $u(x) \in W^{1,2}(\Omega)$, verify by comparing norms:

$$\left(\int_\Omega |u(x)|^6 \, dx \right)^{1/6} \leq C \left(\int_\Omega |u(x)|^2 + |\nabla u(x)|^2 \, dx \right)^{1/2}.$$

   This illustrates the embedding holds with a constant C.

Practice Problems 2

1. Show that the weak derivative of the absolute value function $u(x) = |x|$ exists and find it.

2. Verify the Poincaré inequality for the function $u(x) = x^2 - x$ in the interval $\Omega = (0, 1)$.

3. Determine if the function $f(x) = \frac{1}{x}$ for $x \in (0, 1)$ is in the Sobolev space $W^{1,1}(0, 1)$.

4. Prove the embedding $W^{1,2}(\Omega) \hookrightarrow L^2(\Omega)$ using the Sobolev embedding theorem, for $\Omega \subset \mathbb{R}^n$.

5. Compute the Sobolev norm $\|u\|_{W^{1,2}(-1,1)}$ for the function $u(x) = x^3$.

6. Apply the extension theorem to extend the function $h(x) = x^2$ defined on $\Omega = (-1, 1)$ to a function $\tilde{h}$ on $\mathbb{R}$, ensuring that the Sobolev norm inequality is satisfied.

Answers 2

1. Show that the weak derivative of the absolute value function $u(x) = |x|$ exists and find it. **Solution:**
   The function $u(x) = |x|$ is piecewise defined as:

   $$u(x) = \begin{cases} x & \text{if } x \geq 0, \\ -x & \text{if } x < 0. \end{cases}$$

   The weak derivative $v(x)$ satisfies:

   $$\int_{-\infty}^{\infty} u(x) \frac{d\phi}{dx} \, dx = - \int_{-\infty}^{\infty} v(x) \phi(x) \, dx$$

   for all test functions $\phi \in C_c^\infty(\mathbb{R})$.

   For $x \geq 0$, $u(x) = x$ with a classical derivative of 1. For $x < 0$, $u(x) = -x$ has a classical derivative of -1. Thus, the weak derivative is:

   $$v(x) = \begin{cases} 1 & \text{if } x > 0, \\ -1 & \text{if } x < 0. \end{cases}$$

92

2. Verify the Poincaré inequality for the function $u(x) = x^2 - x$ in the interval $\Omega = (0,1)$. **Solution:**

   The Poincaré inequality states:
   $$\|u\|_{L^p(0,1)} \leq C\|\nabla u\|_{L^p(0,1)}.$$

   For $u(x) = x^2 - x$, calculate:
   $$\nabla u(x) = \frac{du}{dx} = 2x - 1.$$

   Compute:
   $$\|u\|_{L^2(0,1)} = \left(\int_0^1 (x^2 - x)^2 \, dx\right)^{1/2}$$

   and
   $$\|\nabla u\|_{L^2(0,1)} = \left(\int_0^1 (2x - 1)^2 \, dx\right)^{1/2}.$$

   Checking calculations:
   $$\int_0^1 (x^2 - x)^2 \, dx = \int_0^1 (x^4 - 2x^3 + x^2) \, dx = \left[\frac{x^5}{5} - \frac{x^4}{2} + \frac{x^3}{3}\right]_0^1 = \frac{1}{30}.$$

   $$\int_0^1 (2x - 1)^2 \, dx = \int_0^1 (4x^2 - 4x + 1) \, dx = \left[\frac{4x^3}{3} - 2x^2 + x\right]_0^1 = \frac{1}{3}.$$

   Thus:
   $$\left(\frac{1}{30}\right)^{1/2} \leq C\left(\frac{1}{3}\right)^{1/2}, \quad \text{with } C = \sqrt{10}$$

3. Determine if the function $f(x) = \frac{1}{x}$ for $x \in (0,1)$ is in the Sobolev space $W^{1,1}(0,1)$. **Solution:**

   A function $f \in W^{1,1}(0,1)$ requires both f and its weak derivative to be in $L^1(0,1)$.

   Calculating:
   $$\int_0^1 \left|\frac{1}{x}\right| \, dx = \int_0^1 \frac{1}{x} \, dx = \lim_{\epsilon \to 0^+} [\ln(x)]_\epsilon^1 = -\infty.$$

   Since the integral diverges, $f(x) = \frac{1}{x}$ is not in $L^1(0,1)$ and therefore not in $W^{1,1}(0,1)$.

4. Prove the embedding $W^{1,2}(\Omega) \hookrightarrow L^2(\Omega)$ using the Sobolev embedding theorem, for $\Omega \subset \mathbb{R}^n$. **Solution:**

   The Sobolev embedding theorem gives conditions for embedding Sobolev spaces into Lebesgue spaces. For $k = 1$ and $p = 2$, if $pk < n$, then:
   $$W^{k,p}(\Omega) \hookrightarrow L^q(\Omega), \quad q = \frac{np}{n - pk}.$$

   For $\Omega \subset \mathbb{R}^n$ and appropriate dimensions with $pk \geq n$, the embedding simplifies to $L^2(\Omega)$.

   Verify the case $n = 1, p = 2, k = 1$:
   $$q = \frac{2 \cdot 1}{1 - 1 \cdot 1} \to \text{undefined, implies necessity of transformation space.}$$

   For correct dimensional assumptions, identity directly results:
   $$W^{1,2}(\Omega) \hookrightarrow L^2(\Omega).$$

5. Compute the Sobolev norm $\|u\|_{W^{1,2}(-1,1)}$ for the function $u(x) = x^3$. **Solution:**

Sobolev norm:

$$\|u\|_{W^{1,2}(-1,1)} = \left(\int_{-1}^{1} |u(x)|^2 \, dx + \int_{-1}^{1} \left| \frac{du}{dx} \right|^2 \, dx \right)^{1/2}.$$

Here:

$$u(x) = x^3, \quad \frac{du}{dx} = 3x^2.$$

Compute:

$$\int_{-1}^{1} (x^3)^2 \, dx = \int_{-1}^{1} x^6 \, dx = \left[\frac{x^7}{7} \right]_{-1}^{1} = \frac{2}{7}.$$

$$\int_{-1}^{1} (3x^2)^2 \, dx = 9 \int_{-1}^{1} x^4 \, dx = 9 \left[\frac{x^5}{5} \right]_{-1}^{1} = \frac{18}{5}.$$

Hence, the Sobolev norm is:

$$\|u\|_{W^{1,2}(-1,1)} = \left(\frac{2}{7} + \frac{18}{5} \right)^{1/2} = \left(\frac{10 + 126}{35} \right)^{1/2} = \left(\frac{136}{35} \right)^{1/2}.$$

6. Apply the extension theorem to extend the function $h(x) = x^2$ defined on $\Omega = (-1, 1)$ to a function $\tilde{h}$ on $\mathbb{R}$, ensuring that the Sobolev norm inequality is satisfied. **Solution:**

Given the extension theorem, for $h \in W^{k,p}(\Omega)$, there exists an extension $\tilde{h} \in W^{k,p}(\mathbb{R})$ such that:

$$\|\tilde{h}\|_{W^{k,p}(\mathbb{R})} \leq C \|h\|_{W^{k,p}(\Omega)}.$$

Define $\tilde{h}(x)$ outside $\Omega = (-1, 1)$ as:

$$\tilde{h}(x) = \begin{cases} x^2 & \text{for } x \in (-1, 1), \\ 0 & \text{for } x \notin (-1, 1). \end{cases}$$

Check Sobolev norm balance:

$$\|\tilde{h}\|_{W^{k,2}(\mathbb{R})} = \|\tilde{h}\|_{L^2(\mathbb{R})} + \|\nabla \tilde{h}\|_{L^2(\mathbb{R})}.$$

Ensure computation within bounds:

$$\|\tilde{h}\|_{L^2(\mathbb{R})} = \int_{-1}^{1} x^4 \, dx = \frac{2}{5}.$$

$$\|\nabla \tilde{h}\|_{L^2(\mathbb{R})} = \int_{-1}^{1} (2x)^2 \, dx = 4 \cdot \frac{2}{3} = \frac{8}{3}.$$

Check inequality with approximated constant $\|h\|_{W^{k,2}(\Omega)} \equiv$ bounded equivalence.

Practice Problems 3

1. Prove that if $u \in W^{1,p}(\Omega)$, where $\Omega \subset \mathbb{R}^n$ is bounded, the embedding $W^{1,p}(\Omega) \hookrightarrow L^q(\Omega)$ is continuous for $1 \leq p < n$ and q such that $\frac{1}{q} = \frac{1}{p} - \frac{1}{n}$.

2. Show that the weak derivative of the function $u(x) = |x|$ on $\mathbb{R}$ is $u'(x) = \text{sgn}(x)$, where $\text{sgn}(x)$ is the sign function.

3. Use the Poincaré inequality to demonstrate that every function $u \in W_0^{1,p}(\Omega)$ has a norm equivalence: $\|u\|_{W^{1,p}(\Omega)} \approx \|\nabla u\|_{L^p(\Omega)}$.

4. Verify the Sobolev inequality for $u \in W^{1,p}(\mathbb{R}^n)$ and show that if $p < n$, then $\|u\|_{L^{p^*}(\mathbb{R}^n)} \leq C\|\nabla u\|_{L^p(\mathbb{R}^n)}$, where $p^* = \frac{np}{n-p}$.

5. Show that any function $u \in W^{k,p}(\Omega)$ has a smooth extension to $\mathbb{R}^n$ if Ω has a Lipschitz boundary.

6. Explain how the use of weak derivatives facilitates the variational formulation of the Laplace equation $-\Delta u = f$, for $f \in L^2(\Omega)$.

Answers 3

1. Prove that if $u \in W^{1,p}(\Omega)$, where $\Omega \subset \mathbb{R}^n$ is bounded, the embedding $W^{1,p}(\Omega) \hookrightarrow L^q(\Omega)$ is continuous for $1 \leq p < n$ and q such that $\frac{1}{q} = \frac{1}{p} - \frac{1}{n}$.

   Solution: By the Sobolev Embedding Theorem, for $1 \leq p < n$, there exists a continuous imbedding $W^{1,p}(\Omega) \hookrightarrow L^q(\Omega)$ where $q = \frac{np}{n-p}$. This is shown by recognizing that the derivatives $\frac{\partial u}{\partial x_i}$ are in $L^p(\Omega)$, and thus, using the continuous embedding factors into the space $L^q(\Omega)$ where $q > p$ is allowed by the inequality of compactness in $\mathbb{R}^n$.

2. Show that the weak derivative of the function $u(x) = |x|$ on $\mathbb{R}$ is $u'(x) = \text{sgn}(x)$, where $\text{sgn}(x)$ is the sign function.

   Solution: For any test function $\phi \in C_c^\infty(\mathbb{R})$, we calculate:
   $$\int_{\mathbb{R}} |x| \phi'(x) \, dx = - \int_{\mathbb{R}} \text{sgn}(x) \phi(x) \, dx.$$

   The integration by parts shows that $\text{sgn}(x)$ serves as the weak derivative of $|x|$.

3. Use the Poincaré inequality to demonstrate that every function $u \in W_0^{1,p}(\Omega)$ has a norm equivalence: $\|u\|_{W^{1,p}(\Omega)} \approx \|\nabla u\|_{L^p(\Omega)}$.

   Solution: The Poincaré inequality implies:
   $$\|u\|_{L^p(\Omega)} \leq C \|\nabla u\|_{L^p(\Omega)}.$$

   Hence, we have:
   $$\|u\|_{W^{1,p}(\Omega)} \leq C \left(\|u\|_{L^p(\Omega)} + \|\nabla u\|_{L^p(\Omega)} \right) \approx \|\nabla u\|_{L^p(\Omega)}.$$

4. Verify the Sobolev inequality for $u \in W^{1,p}(\mathbb{R}^n)$ and show that if $p < n$, then $\|u\|_{L^{p^*}(\mathbb{R}^n)} \leq C \|\nabla u\|_{L^p(\mathbb{R}^n)}$, where $p^* = \frac{np}{n-p}$.

   Solution: The Sobolev inequality gives us:
   $$\|u\|_{L^{p^*}(\mathbb{R}^n)} \leq C \|u\|_{W^{1,p}(\mathbb{R}^n)}.$$

   For $u \in W^{1,p}(\mathbb{R}^n)$, we have:
   $$\|u\|_{L^{p^*}(\mathbb{R}^n)} \leq C \|\nabla u\|_{L^p(\mathbb{R}^n)} \text{ with } p^* = \frac{np}{n-p}.$$

5. Show that any function $u \in W^{k,p}(\Omega)$ has a smooth extension to $\mathbb{R}^n$ if Ω has a Lipschitz boundary.

   Solution: By the extension theorem, a $u \in W^{k,p}(\Omega)$ has a continuous extension $\tilde{u} \in W^{k,p}(\mathbb{R}^n)$ such that:
   $$\|\tilde{u}\|_{W^{k,p}(\mathbb{R}^n)} \leq C \|u\|_{W^{k,p}(\Omega)}.$$

   The regularity of the boundary $\partial \Omega$ allows us to construct such extensions without exceeding the original norm bound.

6. Explain how the use of weak derivatives facilitates the variational formulation of the Laplace equation $-\Delta u = f$, for $f \in L^2(\Omega)$.

Solution: The Laplace equation $-\Delta u = f$ can be reformulated in a weak sense by seeking $u \in H_0^1(\Omega)$ such that:

$$\int_\Omega \nabla u \cdot \nabla v \, dx = \int_\Omega fv \, dx, \quad \forall v \in H_0^1(\Omega).$$

This reformulation captures solutions that might not possess classical derivatives, allowing for the application of modern functional analytic techniques to obtain solution existence and uniqueness.

Chapter 10

Rigged Hilbert Spaces and Gelfand Triplets

Introduction to Rigged Hilbert Spaces

Rigged Hilbert spaces, also known as Gelfand triplets, provide a mathematical framework for handling unbounded operators and generalized eigenvectors. A rigged Hilbert space is typically constructed as a triplet of the form:

$$\Phi \subset \mathcal{H} \subset \Phi',$$

where $\mathcal{H}$ is a Hilbert space, Φ is a dense subspace of $\mathcal{H}$, and Φ' represents the dual space of Φ.

Construction of Gelfand Triplets

To construct a Gelfand triplet, begin with a Hilbert space $\mathcal{H}$. Identify a dense subset $\Phi \subset \mathcal{H}$ such that the topology on Φ is finer than that induced by $\mathcal{H}$. The dual space Φ' then consists of all continuous linear functionals on Φ.

$$(\phi, \psi)_{\mathcal{H}} = \langle \phi, \psi \rangle, \quad \forall \phi \in \Phi, \psi \in \mathcal{H}. \tag{10.1}$$

The triplet $\Phi \subset \mathcal{H} \subset \Phi'$ allows for the consideration of distributions that generalize the notion of vectors in $\mathcal{H}$.

Role in Spectral Decomposition of Unbounded Operators

Rigged Hilbert spaces play a crucial role in the spectral decomposition of unbounded operators, which are common in quantum mechanics. Operators often extend beyond the traditional Hilbert space framework:

$$T : \mathcal{D}(T) \subset \mathcal{H} \to \mathcal{H}, \tag{10.2}$$

where $\mathcal{D}(T)$ is the domain of T, a dense subset in $\mathcal{H}$.

1 The Spectral Theorem

For self-adjoint unbounded operators, the spectral theorem provides a decomposition in terms of projection-valued measures. In a rigged Hilbert space, eigenvectors associated with unbounded operators might not reside in the Hilbert space but can be captured in the distributional sense within Φ'.

Consider the unbounded operator A with eigenvalue equation:

$$A\psi = \lambda\psi, \quad \psi \in \Phi'. \tag{10.3}$$

Here, ψ may be treated as a generalized eigenfunction, where λ is a complex eigenvalue. The spectral theorem then asserts:

$$A = \int \lambda \, dE(\lambda), \tag{10.4}$$

for some spectral measure $E(\lambda)$.

2 Applications to Quantum Mechanics

In quantum mechanics, the observables are represented as operators on Hilbert spaces. Rigged Hilbert spaces provide a framework to deal with operators like the momentum or position operators, which are inherently unbounded. Such a setup allows for a more complete treatment of eigenvalue problems with continuous spectra.

Consider an operator Q in quantum mechanics, representing a physical quantity, possibly unbounded:

$$Q\phi = q\phi, \quad \phi \in \Phi', \tag{10.5}$$

where the spectrum of Q can be decomposed via the spectral theorem into a direct integral form. This decomposition using a rigged Hilbert space allows one to accommodate both discrete and continuous spectral components in a unified theory.

3 Example: The Harmonic Oscillator

As an example, consider the quantum harmonic oscillator with Hamiltonian H:

$$H = -\frac{\hbar^2}{2m}\Delta + \frac{1}{2}m\omega^2 x^2. \tag{10.6}$$

Its eigenfunctions form a complete set in $\mathcal{H}$, yet in the rigged space, one introduces generalized solutions corresponding to distributional eigenstates, facilitating the use of coherent states or other unconventional bases.

Practice Problems 1

1. Describe the construction of a Gelfand triplet starting from a given Hilbert space $\mathcal{H}$. Define each of the spaces Φ and Φ' in the triplet:
$$\Phi \subset \mathcal{H} \subset \Phi'.$$

2. Explain how rigged Hilbert spaces facilitate the spectral decomposition of unbounded operators in quantum mechanics. What is the role of generalized eigenvectors in Φ'?

3. Consider an unbounded self-adjoint operator A on a Hilbert space $\mathcal{H}$. Demonstrate how the spectral theorem applies within the context of a rigged Hilbert space.

4. Provide an example of an operator in quantum mechanics that benefits from the framework of rigged Hilbert spaces for a complete treatment. Discuss the significance of the continuous spectrum in this context.

5. Describe how the quantum harmonic oscillator can be understood using rigged Hilbert spaces, particularly focusing on distributional eigenstates.

6. Discuss the implications of using rigged Hilbert spaces for physical observables with continuous spectra, like the position operator.

Answers 1

1. **Describe the construction of a Gelfand triplet starting from a given Hilbert space $\mathcal{H}$.**

Solution:

Start with a Hilbert space $\mathcal{H}$. Define a dense subspace $\Phi \subset \mathcal{H}$ with a topology that is finer than the one induced by $\mathcal{H}$. This means that convergence in Φ guarantees convergence in $\mathcal{H}$, but not necessarily vice versa.

The dual space Φ' is then defined as the set of all continuous linear functionals on Φ. These functionals can be understood as 'generalized vectors' in Φ', completing the triplet $\Phi \subset \mathcal{H} \subset \Phi'$. This framework allows operators and vectors that are not well-defined on $\mathcal{H}$ alone but are significant in quantum theory.

2. **Explain how rigged Hilbert spaces facilitate the spectral decomposition of unbounded operators in quantum mechanics.**

   Solution:

   In quantum mechanics, unbounded operators such as position and momentum are common. Within a Hilbert space, these operators can pose difficulties, as their eigenvectors do not belong to the space. Rigged Hilbert spaces address this by extending the notion of eigenvectors to encompass generalized eigenvectors located in the dual space Φ'.

   The spectral theorem, when applied in this broader context, allows us to decompose unbounded operators using projection-valued measures. Generalized eigenvectors corresponding to eigenvalues of the operator can be captured in Φ', accommodating both discrete and continuous spectra.

3. **Consider an unbounded self-adjoint operator A on a Hilbert space $\mathcal{H}$. Demonstrate how the spectral theorem applies within the context of a rigged Hilbert space.**

   Solution:

   For a self-adjoint operator A, the spectral theorem in a rigged Hilbert space framework asserts that:

   $$A = \int \lambda \, dE(\lambda),$$

   where $E(\lambda)$ is a spectral measure. In the context of $\Phi' \subset \mathcal{H}$, each λ corresponds to generalized eigenvectors in Φ'. This extends the concept of the spectrum to include continuous parts and allows for a complete integration over the entire spectrum, thus achieving the operator's decomposition in terms of its spectral components.

4. **Provide an example of an operator in quantum mechanics that benefits from the framework of rigged Hilbert spaces for a complete treatment.**

   Solution:

   The momentum operator in quantum mechanics is inherently unbounded and its eigenfunctions, exponential functions, do not belong to the typical Hilbert space $L^2(\mathbb{R})$. The rigged Hilbert space framework allows the momentum operator's deficiencies in a pure L^2 framework to be resolved by extending the space to Φ', wherein generalized eigenvectors and continuous spectra are naturally incorporated.

5. **Describe how the quantum harmonic oscillator can be understood using rigged Hilbert spaces.**

   Solution:

   The quantum harmonic oscillator's Hamiltonian has a complete set of eigenfunctions in a Hilbert space, but to understand coherent states and other states that are not plain L^2 functions, rigged Hilbert spaces are employed. In Φ', solutions can be treated as distributions, facilitating the use of broader bases that reveal richer physical phenomena, such as coherent and squeezed states.

6. **Discuss the implications of using rigged Hilbert spaces for physical observables with continuous spectra, like the position operator.**

   Solution:

Observables like the position operator can have continuous spectra, making traditional eigenvalue solutions inadequate. The rigged Hilbert space approach enables the construction of generalized eigenvectors corresponding to every real number, λ, in Φ', where λ represents possible measurement outcomes. This allows for a precise treatment and proper spectral decomposition of such operators, crucial for accurate quantum mechanical analysis.

Practice Problems 2

1. Define a rigged Hilbert space and explain how it differs from a traditional Hilbert space.

2. Describe the role of the Gelfand triplet in accommodating unbounded operators.

3. Explain how the spectral decomposition is applied to unbounded operators in a rigged Hilbert space.

4. Provide an example of a physical observable in quantum mechanics that is best analyzed using a rigged Hilbert space.

5. Demonstrate with an example how generalized eigenvectors are utilized within the framework of rigged Hilbert spaces.

6. Using the example of the harmonic oscillator, explain the benefits of employing a rigged Hilbert space over a traditional Hilbert space.

Answers 2

1. Define a rigged Hilbert space and explain how it differs from a traditional Hilbert space. **Solution:** A rigged Hilbert space, or Gelfand triplet, is a triplet of spaces:

$$\Phi \subset \mathcal{H} \subset \Phi',$$

where $\mathcal{H}$ is a Hilbert space, Φ is a dense subspace with a finer topology, and Φ' is the dual of Φ. Unlike traditional Hilbert spaces, rigged Hilbert spaces can accommodate distributional (generalized) eigenvectors and expand the scope of spectral analysis beyond bounded operators.

2. Describe the role of the Gelfand triplet in accommodating unbounded operators. **Solution:** The Gelfand triplet provides a framework in which unbounded operators, common in quantum mechanics, can be treated more effectively. In a traditional Hilbert space, eigenvectors for unbounded operators may not exist as elements of the space. The inclusion of Φ', the dual of a dense subspace Φ, allows these operators to have generalized eigenvectors, giving a broader interpretation of spectral decomposition.

3. Explain how the spectral decomposition is applied to unbounded operators in a rigged Hilbert space. **Solution:** In a rigged Hilbert space, spectral decomposition involves using projection-valued measures to express unbounded operators as integrals over their spectra. The generalized eigenvectors in Φ' are used to represent eigenfunctions not present in $\mathcal{H}$, thereby enabling the inclusion of continuous spectra. The spectral theorem in this context asserts that for any self-adjoint operator, an integral form involving the spectral measure allows for a complete description even outside the confines of the traditional Hilbert space.

4. Provide an example of a physical observable in quantum mechanics that is best analyzed using a rigged Hilbert space. **Solution:** The momentum operator, commonly denoted as $\hat{p} = -i\hbar\frac{d}{dx}$, serves as an example. It is unbounded and generally not defined on the entire Hilbert space $L^2(\mathbb{R})$. As such, it requires a rigged Hilbert space to fully develop its spectral properties and include its continuous spectrum.

5. Demonstrate with an example how generalized eigenvectors are utilized within the framework of rigged Hilbert spaces. **Solution:** Consider the position operator Q, which is typically unbounded. Generalized eigenvectors may be represented using Dirac delta functions, $\delta(x - x_0)$, indicating the position x_0 in space. In a standard Hilbert space, such functions do not qualify as vectors; however, they fit within Φ', enabling their use in quantum mechanics to describe states with definite values.

6. Using the example of the harmonic oscillator, explain the benefits of employing a rigged Hilbert space over a traditional Hilbert space. **Solution:** For the quantum harmonic oscillator, the Hamiltonian $H = -\frac{\hbar^2}{2m}\Delta + \frac{1}{2}m\omega^2 x^2$ can be analyzed in a rigged Hilbert space to explore both bound states and the richer structure of the system including distributional states. While the ensuring basis in $\mathcal{H}$ adheres to standard normalizable solutions, in Φ', one can incorporate coherent states and other non-standard bases better suited to applications in quantum optics and other advanced quantum scenarios.

Practice Problems 3

1. Define a Gelfand triplet and explain its role in the context of quantum mechanics. 4cm

2. Consider a Hilbert space $\mathcal{H}$ with a dense subset Φ. Show how the dual space Φ' can be constructed and interpreted. 4cm

3. Discuss why rigged Hilbert spaces are necessary for handling unbounded operators in quantum mechanics. 4cm

4. Prove that any self-adjoint operator on a dense domain $\mathcal{D}(T)$ in a Hilbert space $\mathcal{H}$ has a spectral decomposition involving spectral measures. 4cm

5. Consider the Hamiltonian operator $H = -\frac{\hbar^2}{2m}\Delta + \frac{1}{2}m\omega^2 x^2$. Describe how rigged Hilbert spaces facilitate solutions beyond square-integrable eigenfunctions. 4cm

6. Explain how Gelfand triplets extend the applicability of the spectral theorem to operators with continuous spectra in quantum mechanics. 4cm

Answers 3

1. **Define a Gelfand triplet and explain its role in the context of quantum mechanics.**

   Solution: A Gelfand triplet $\Phi \subset \mathcal{H} \subset \Phi'$ incorporates a Hilbert space $\mathcal{H}$, a dense subspace Φ with finer topology, and its dual Φ'. This framework extends the Hilbert space approach to handle generalized eigenvectors, which are crucially needed in quantum mechanics. Such a triplet allows us to rigorously treat observables represented by unbounded operators, extending the orthodox treatment to encompass eigenstates not residing in $\mathcal{H}$ and facilitating a more comprehensive analysis of quantum states.

2. **Consider a Hilbert space $\mathcal{H}$ with a dense subset Φ. Show how the dual space Φ' can be constructed and interpreted.**

   Solution: The dense subset $\Phi \subset \mathcal{H}$ is assigned a topology finer than $\mathcal{H}$'s, generally through a sequence of norms or seminorms making it a locally convex space. The dual space Φ' consists of continuous linear functionals on Φ. For any $\phi \in \Phi$ and $\psi \in \Phi'$, we compute $\psi(\phi)$ using these dual operations, extending the inner product space $(\phi, \psi)_{\mathcal{H}} = \langle \phi, \psi \rangle$. This allows interpretation of elements in Φ' as generalized vectors which extend the concept of quantum states beyond $\mathcal{H}$.

3. **Discuss why rigged Hilbert spaces are necessary for handling unbounded operators in quantum mechanics.**

   Solution: Rigged Hilbert spaces address the inadequacy of ordinary Hilbert spaces to accommodate unbounded operators and their eigenfunctions. Many operators in quantum mechanics, such as momentum and position, are unbounded and their spectra include both discrete and continuous parts.

Traditional Hilbert spaces only manage bounded operators or subsets of eigenfunctions. Rigged Hilbert spaces, incorporating generalized functions (elements of Φ'), allow a rigorous treatment of all eigenstates and ensure all operatorial operations, including spectral decomposition, remain well-defined.

4. **Prove that any self-adjoint operator on a dense domain $\mathcal{D}(T)$ in a Hilbert space $\mathcal{H}$ has a spectral decomposition involving spectral measures.**

   Solution: Consider a self-adjoint operator T defined on $\mathcal{D}(T) \subset \mathcal{H}$. The spectral theorem guarantees that if $T = T^*$, it admits a projection-valued measure E such that:

   $$T = \int \lambda \, dE(\lambda),$$

   on $\mathcal{D}(T)$. This decomposition reflects that T can be expressed in terms of its eigenvalues integrated over the spectrum, represented by E. In rigged spaces, the inclusion of generalized eigenvectors in Φ' means even unbounded operators with continuous spectra can abide by this spectral formulation.

5. **Consider the Hamiltonian operator $H = -\frac{\hbar^2}{2m}\Delta + \frac{1}{2}m\omega^2 x^2$. Describe how rigged Hilbert spaces facilitate solutions beyond square-integrable eigenfunctions.**

   Solution: The Hamiltonian of a quantum harmonic oscillator often results in solutions that are not strictly within $\mathcal{H}$, especially when considering coherent or other generalized states. Within the rigged Hilbert framework, one defines Φ to be the Schwartz space (rapidly decreasing functions), allowing elements of Φ' to encompass distributions, such as the delta function, expanding the eigenfunction basis. These generalized eigenfunctions facilitate solution representation as generalized series or distributional bases for solving quantum problems that extend beyond conventional eigenvectors.

6. **Explain how Gelfand triplets extend the applicability of the spectral theorem to operators with continuous spectra in quantum mechanics.**

   Solution: Gelfand triplets allow the spectral theorem to remain valid for operators whose spectra feature non-discreteness, unlike conventional Hilbert spaces which rely on orthogonality of square-integrable eigenfunctions. In a rigged context, the triplet $\Phi \subset \mathcal{H} \subset \Phi'$ supports continuous spectral components—eigenstates may reside in Φ' and need not converge in the metric of $\mathcal{H}$. This ability to rigorously derive spectral decompositions in the wider context of Φ and Φ' makes Gelfand triplets indispensable for formulating quantum mechanics of observables like momentum, where continuous spectrums are predicted.

Chapter 11

Introduction to Quantum Mechanics in Hilbert Space

Quantum States as Vectors

In quantum mechanics, the state of a physical system is represented by a vector in a complex Hilbert space $\mathcal{H}$. A ket ψ in the Dirac notation denotes such a state. For a normalized state, the vector satisfies:

$$\langle \psi | \psi \rangle = 1,$$

where $\langle \psi | \psi \rangle$ is the inner product in $\mathcal{H}$.

1 Superposition Principle

If ψ_1 and ψ_2 are valid states, their linear combination $\alpha \psi_1 + \beta \psi_2$ for complex numbers α and β also represents a valid state, emphasizing the principle of superposition.

2 Orthogonality and Orthonormality

Two states ϕ and ψ are orthogonal if:

$$\langle \phi | \psi \rangle = 0.$$

An orthonormal set $\{\psi_n\}$ satisfies $\langle \psi_m | \psi_n \rangle = \delta_{mn}$, where δ_{mn} is the Kronecker delta.

Operators as Observables

In quantum mechanics, physical observables correspond to linear operators on $\mathcal{H}$. An observable A is represented by an operator $\hat{A}$, with observables defined as self-adjoint operators.

$$\hat{A} = \hat{A}^\dagger, \tag{11.1}$$

where $\hat{A}^\dagger$ is the adjoint operator.

1 Eigenvalues and Eigenvectors

The measurement result of an observable is one of the eigenvalues of the corresponding operator. If $\hat{A}\psi = \lambda \psi$, λ is the eigenvalue and ψ the associated eigenvector.

2 Projection Operators

Projection operators P are important in measurement postulates, defined by:

$$P^2 = P, \quad P^\dagger = P.$$

They map vectors to subspaces associated with definite measurement outcomes.

Probabilistic Interpretation

In this framework, quantum mechanics introduces a probabilistic nature to measurement outcomes.

1 Born Rule

The probability of obtaining a measurement value λ for an observable A when the system is in state ψ is given by:

$$P(\lambda) = \|\phi\psi\|^2,$$

where ϕ is an eigenvector of $\hat{A}$ associated with λ.

2 Expectation Values

The expectation value $\langle \hat{A} \rangle$ of an observable $\hat{A}$ in state ψ is:

$$\langle \hat{A} \rangle = \langle \psi | \hat{A} | \psi \rangle,$$

providing the average value over many measurements.

State Evolution

1 Schrödinger Equation

The time evolution of quantum states is governed by the Schrödinger equation. Given the Hamiltonian $\hat{H}$, the state $\psi(t)$ evolves as:

$$i\hbar \frac{\partial}{\partial t} \psi(t) = \hat{H}\psi(t),$$

where $\hbar$ is the reduced Planck's constant.

2 Unitary Evolution

The evolution is unitary, preserving state norms. If ψ_0 is the initial state, then:

$$\psi(t) = U(t)\psi_0,$$

where $U(t)$ is a unitary operator satisfying $U(t)^\dagger U(t) = I$.

Measurement Postulate

Upon measurement, the state undergoes a collapse to an eigenstate of the observable. If $\hat{A}$ is measured and ϕ is its eigenstate, the system collapses to:

$$\frac{P\psi}{\|P\psi\|},$$

where P is the projection onto the subspace spanned by ϕ.

Example: Spin-$\frac{1}{2}$ Systems

Consider a spin-$\frac{1}{2}$ particle, described by a two-dimensional Hilbert space. The state is given by:

$$\psi = \alpha\uparrow + \beta\downarrow,$$

where α and β are complex numbers satisfying $|\alpha|^2 + |\beta|^2 = 1$.

1   Spin Operators

The spin operators S_x, S_y, S_z are represented as matrices:

$$S_z = \frac{\hbar}{2}\begin{pmatrix} 1 & 0 \\ 0 & -1 \end{pmatrix}, \quad S_x = \frac{\hbar}{2}\begin{pmatrix} 0 & 1 \\ 1 & 0 \end{pmatrix},$$

acting on the spin state vector.

2   Measurement and Probabilities

Measurement of S_z yields $\frac{\hbar}{2}$ with probability $|\alpha|^2$ or $-\frac{\hbar}{2}$ with probability $|\beta|^2$, illustrating the probabilistic nature of quantum mechanics.

Practice Problems 1

1. Verify the normalization condition for a quantum state $\psi = \sqrt{\frac{1}{3}}\phi_1 + \sqrt{\frac{2}{3}}\phi_2$, where ϕ_1 and ϕ_2 are orthonormal vectors.

2. Prove that if ψ is an eigenvector of a self-adjoint operator $\hat{A}$ with eigenvalue λ, then λ is real.

3. Demonstrate that the set $\{\chi_1, \chi_2\}$, where $\chi_1 = \frac{1}{\sqrt{2}}(\phi_1 + \phi_2)$ and $\chi_2 = \frac{1}{\sqrt{2}}(\phi_1 - \phi_2)$, is orthonormal given that $\{\phi_1, \phi_2\}$ is orthonormal.

4. Calculate the expectation value of the spin operator S_z for a spin-$\frac{1}{2}$ particle in the state $\psi = \alpha\uparrow + \beta\downarrow$ with $S_z = \frac{\hbar}{2}\begin{pmatrix} 1 & 0 \\ 0 & -1 \end{pmatrix}$.

5. Show that the time evolution operator $U(t) = e^{-i\hat{H}t/\hbar}$ is unitary if $\hat{H}$ is a self-adjoint operator (Hamiltonian).

6. For the projection operator P onto a subspace V spanned by an orthonormal set $\{v_i\}$, prove that $P^2 = P$.

Answers 1

1. Verify the normalization condition:
   Solution:
   $$\langle\psi|\psi\rangle = \left(\sqrt{\frac{1}{3}}\phi_1 + \sqrt{\frac{2}{3}}\phi_2\right)\left(\sqrt{\frac{1}{3}}\phi_1 + \sqrt{\frac{2}{3}}\phi_2\right)$$
   $$= \frac{1}{3}\langle\phi_1|\phi_1\rangle + \frac{2}{3}\langle\phi_2|\phi_2\rangle + \sqrt{\frac{2}{9}}(\langle\phi_1|\phi_2\rangle + \langle\phi_2|\phi_1\rangle)$$
   $$= \frac{1}{3} + \frac{2}{3} + 0 = 1$$
   Thus, ψ is normalized.

2. Prove eigenvalue reality:
   Solution:
   $$\langle\psi|\hat{A}\psi\rangle = \langle\psi|\lambda\psi\rangle = \lambda\langle\psi|\psi\rangle = \lambda$$

Because $\hat{A}$ is self-adjoint:

$$\langle\psi|\hat{A}\psi\rangle = \langle\hat{A}\psi|\psi\rangle = \overline{\langle\psi|\hat{A}\psi\rangle} = \overline{\lambda}$$

Therefore, $\lambda = \overline{\lambda}$ which means λ is real.

3. Orthogonality of given vectors:
   Solution:

$$\langle\chi_1|\chi_2\rangle = \frac{1}{2}((\langle\phi_1| + \langle\phi_2|)(\phi_1 - \phi_2)$$

$$= \frac{1}{2}(\langle\phi_1|\phi_1\rangle - \langle\phi_1|\phi_2\rangle + \langle\phi_2|\phi_1\rangle - \langle\phi_2|\phi_2\rangle)$$

$$= \frac{1}{2}(1 - 0 + 0 - 1) = 0$$

Thus, $\{\chi_1, \chi_2\}$ is orthonormal.

4. Calculate expectation value:
   Solution:

$$\langle\hat{S}_z\rangle = \langle\psi|\hat{S}_z|\psi\rangle$$

$$= (\alpha^*{\uparrow} + \beta^*{\downarrow})\left(\frac{\hbar}{2}\begin{pmatrix} 1 & 0 \\ 0 & -1 \end{pmatrix}\right)(\alpha{\uparrow} + \beta{\downarrow})$$

$$= \frac{\hbar}{2}(|\alpha|^2 - |\beta|^2)$$

5. Unitarity of evolution operator:
   Solution:

$$U(t)^{\dagger} = \left(e^{-i\hat{H}t/\hbar}\right)^{\dagger} = e^{i\hat{H}t/\hbar}$$

$$U(t)^{\dagger}U(t) = e^{i\hat{H}t/\hbar}e^{-i\hat{H}t/\hbar} = I$$

This proves that $U(t)$ is unitary.

6. Proof for projection:
   Solution:

$$P = \sum v_i v_i$$

$$P^2 = \left(\sum v_i v_i\right)\left(\sum v_j v_j\right)$$

$$= \sum\sum v_i(v_i|v_j)v_j = \sum v_i v_i = P$$

Hence, $P^2 = P$.

Practice Problems 2

1. Consider a quantum state ψ in a Hilbert space $\mathcal{H}$. Show that if ψ is normalized, then $\psi\psi = 1$.

2. Let ϕ and ψ be orthogonal quantum states in a Hilbert space. Verify the orthogonality condition with respect to the inner product.

3. Show that for two quantum states ψ_1 and ψ_2, their superposition $\alpha\psi_1 + \beta\psi_2$ is valid with the normalization condition $|\alpha|^2 + |\beta|^2 = 1$.

4. Given the Hamiltonian operator $\hat{H}$, derive the time-dependent Schrödinger equation for a quantum state ψ.

5. Consider the spin operators S_x and S_z for a spin-$\frac{1}{2}$ system. Compute the commutator $[S_x, S_z]$.

6. Calculate the probability of measuring the state $\uparrow$ for a spin-$\frac{1}{2}$ particle in the superposition state $\psi = \frac{1}{\sqrt{2}}\uparrow + \frac{1}{\sqrt{2}}\downarrow$.

Answers 2

1. Consider a quantum state ψ in a Hilbert space $\mathcal{H}$. Show that if ψ is normalized, then $\psi\psi = 1$.

   Solution: The normalization of a state ψ is given by the condition:

   $$\psi\psi = 1.$$

   Since normalization implies that the probability of the state existing is unity, the inner product $\psi\psi$ inherently equals one for a normalized state.

2. Let ϕ and ψ be orthogonal quantum states in a Hilbert space. Verify the orthogonality condition with respect to the inner product.

   Solution: Orthogonality between two states is defined by the inner product:

   $$\phi\psi = 0.$$

   Verifying this condition involves computing the inner product and confirming that it equals zero, indicating no overlap between the states in the space.

3. Show that for two quantum states ψ_1 and ψ_2, their superposition $\alpha\psi_1 + \beta\psi_2$ is valid with the normalization condition $|\alpha|^2 + |\beta|^2 = 1$.

   Solution: The superposition state is given by:

   $$\Psi = \alpha\psi_1 + \beta\psi_2.$$

   The normalization condition is:

   $$\Psi\Psi = |\alpha|^2\psi_1\psi_1 + |\beta|^2\psi_2\psi_2 + \alpha\overline{\beta}\psi_1\psi_2 + \overline{\alpha}\beta\psi_2\psi_1.$$

   Assuming ψ_1 and ψ_2 are normalized and orthogonal:

   $$\Psi\Psi = |\alpha|^2 + |\beta|^2 = 1.$$

   Thus confirming the superposition is valid and normalized.

4. Given the Hamiltonian operator $\hat{H}$, derive the time-dependent Schrödinger equation for a quantum state ψ.

   Solution: The time-dependent Schrödinger equation is expressed as:

   $$i\hbar\frac{\partial}{\partial t}\psi(t) = \hat{H}\psi(t).$$

   Where the operator $\hat{H}$ is the Hamiltonian, describing the total energy of the quantum system. This equation describes how the quantum state ψ evolves over time.

5. Consider the spin operators S_x and S_z for a spin-$\frac{1}{2}$ system. Compute the commutator $[S_x, S_z]$.

   Solution: The commutator of the two operators is calculated as:

   $$[S_x, S_z] = S_x S_z - S_z S_x.$$

   For spin-$\frac{1}{2}$ operators represented as:

   $$S_x = \frac{\hbar}{2}\begin{pmatrix} 0 & 1 \\ 1 & 0 \end{pmatrix}, \quad S_z = \frac{\hbar}{2}\begin{pmatrix} 1 & 0 \\ 0 & -1 \end{pmatrix},$$

   the product calculations followed by subtraction yield:

   $$[S_x, S_z] = i\hbar S_y.$$

112

6. Calculate the probability of measuring the state $\uparrow$ for a spin-$\frac{1}{2}$ particle in the superposition state $\psi = \frac{1}{\sqrt{2}}\uparrow + \frac{1}{\sqrt{2}}\downarrow$.

   Solution: The probability of measuring $\uparrow$ is given by the Born rule:

   $$P(\uparrow) = |\uparrow\psi|^2.$$

   Expanding the superposition state:

   $$\psi = \frac{1}{\sqrt{2}}\uparrow + \frac{1}{\sqrt{2}}\downarrow.$$

   Therefore, the inner product is:

   $$\uparrow\psi = \frac{1}{\sqrt{2}},$$

   and consequently, the probability is:

   $$P(\uparrow) = \left|\frac{1}{\sqrt{2}}\right|^2 = \frac{1}{2}.$$

   Thus, the probability of measuring the spin state $\uparrow$ is $\frac{1}{2}$.

Practice Problems 3

1. Prove that the inner product $\langle \phi | \psi \rangle$ is linear in its second argument.

2. Given a normalized state $\psi = \alpha\uparrow + \beta\downarrow$ for a spin-$\frac{1}{2}$ system, find the expectation value of the S_z operator.

3. Let $\hat{A}$ be a self-adjoint operator with eigenvectors $\{a_n\}$. Show that the eigenvectors corresponding to distinct eigenvalues are orthogonal.

4. For a quantum state $\psi(t)$ evolving according to the Schrödinger equation with Hamiltonian $\hat{H}$, prove that the norm $\|\psi(t)\|$ is conserved.

5. A projection operator P satisfies $P^2 = P$. Show that P can have only eigenvalues 0 or 1.

6. Calculate the probability of measuring $S_x = \frac{\hbar}{2}$ for a spin-$\frac{1}{2}$ state $\psi = \alpha{\uparrow} + \beta{\downarrow}$.

Answers 3

1. **Prove that the inner product $\langle\phi|\psi\rangle$ is linear in its second argument.**

   Solution: The inner product is linear in its second argument if for any states ψ_1, ψ_2 and scalars α, β, the following holds:
   $$\langle\phi|(\alpha\psi_1 + \beta\psi_2)\rangle = \alpha\langle\phi|\psi_1\rangle + \beta\langle\phi|\psi_2\rangle.$$
   By the definition of linearity, expand the left side:
   $$\langle\phi|(\alpha\psi_1 + \beta\psi_2)\rangle = \langle\phi|\alpha\psi_1\rangle + \langle\phi|\beta\psi_2\rangle,$$
   $$= \alpha\langle\phi|\psi_1\rangle + \beta\langle\phi|\psi_2\rangle.$$
   Thus, linearity is verified.

2. **Given a normalized state $\psi = \alpha{\uparrow} + \beta{\downarrow}$ for a spin-$\frac{1}{2}$ system, find the expectation value of the S_z operator.**

   Solution: The spin operator S_z is given by:
   $$S_z = \frac{\hbar}{2}\begin{pmatrix} 1 & 0 \\ 0 & -1 \end{pmatrix}.$$

114

The expectation value is:
$$\langle S_z \rangle = \langle \psi | S_z | \psi \rangle.$$
Substitute ψ:
$$\langle S_z \rangle = (\alpha^* \uparrow + \beta^* \downarrow) S_z (\alpha \uparrow + \beta \downarrow).$$
Compute terms:
$$= \alpha^* \alpha \langle \uparrow | S_z | \uparrow \rangle + \alpha^* \beta \langle \uparrow | S_z | \downarrow \rangle + \beta^* \alpha \langle \downarrow | S_z | \uparrow \rangle + \beta^* \beta \langle \downarrow | S_z | \downarrow \rangle,$$
$$= \frac{\hbar}{2} (\alpha^* \alpha \cdot 1 + \beta^* \beta \cdot (-1)) = \frac{\hbar}{2} (|\alpha|^2 - |\beta|^2).$$
Thus, $\langle S_z \rangle = \frac{\hbar}{2} (|\alpha|^2 - |\beta|^2)$.

3. **Let $\hat{A}$ be a self-adjoint operator with eigenvectors $\{a_n\}$. Show that the eigenvectors corresponding to distinct eigenvalues are orthogonal.**

   Solution: Let $\hat{A} a_n = \lambda_n a_n$ and $\hat{A} a_m = \lambda_m a_m$ with $\lambda_n \neq \lambda_m$. Consider:
   $$\langle a_m | \hat{A} | a_n \rangle = \lambda_n \langle a_m | a_n \rangle$$
   $$\langle \hat{A} a_m | a_n \rangle = \lambda_m \langle a_m | a_n \rangle.$$
   Since $\hat{A}$ is self-adjoint, $\langle a_m | \hat{A} | a_n \rangle = \langle \hat{A} a_m | a_n \rangle$:
   $$\lambda_n \langle a_m | a_n \rangle = \lambda_m \langle a_m | a_n \rangle.$$
   $$(\lambda_n - \lambda_m) \langle a_m | a_n \rangle = 0.$$
   Since $\lambda_n \neq \lambda_m$, $\langle a_m | a_n \rangle = 0$. Thus, a_n and a_m are orthogonal.

4. **For a quantum state $\psi(t)$ evolving according to the Schrödinger equation with Hamiltonian $\hat{H}$, prove that the norm $\|\psi(t)\|$ is conserved.**

   Solution: The Schrödinger equation is:
   $$i\hbar \frac{\partial}{\partial t} \psi(t) = \hat{H} \psi(t).$$
   Consider the norm squared:
   $$\frac{d}{dt} \langle \psi(t) | \psi(t) \rangle = \langle \frac{d}{dt} \psi(t) | \psi(t) \rangle + \langle \psi(t) | \frac{d}{dt} \psi(t) \rangle.$$
   Substitute from the Schrödinger equation:
   $$= \frac{1}{i\hbar} \langle \hat{H} \psi(t) | \psi(t) \rangle - \frac{1}{i\hbar} \langle \psi(t) | \hat{H} \psi(t) \rangle.$$
   Since $\hat{H}$ is self-adjoint, the two terms are conjugates, giving zero:
   $$= 0,$$
   proving that $\|\psi(t)\|$ is conserved.

5. **A projection operator P satisfies $P^2 = P$. Show that P can have only eigenvalues 0 or 1.**

   Solution: Let $Pv = \lambda v$ for eigenvalue λ. Then:
   $$P^2 v = \lambda^2 v.$$
   But $P^2 = P$, thus:
   $$P^2 v = Pv = \lambda v.$$
   Equating gives:
   $$\lambda^2 = \lambda.$$
   Solving yields $\lambda(\lambda - 1) = 0$, thus $\lambda = 0$ or $\lambda = 1$.

115

6. **Calculate the probability of measuring $S_x = \frac{\hbar}{2}$ for a spin-$\frac{1}{2}$ state $\psi = \alpha{\uparrow} + \beta{\downarrow}$.**

   Solution: The eigenstates of S_x are:

$$+_x = \frac{1}{\sqrt{2}}({\uparrow} + {\downarrow}), \quad -_x = \frac{1}{\sqrt{2}}({\uparrow} - {\downarrow}).$$

The probability P of measuring $\frac{\hbar}{2}$ is:

$$P = |\langle +_x | \psi \rangle|^2.$$

Compute $\langle +_x | \psi \rangle$:

$$\langle +_x | \psi \rangle = \frac{1}{\sqrt{2}}({\uparrow} + {\downarrow})(\alpha{\uparrow} + \beta{\downarrow}),$$

$$= \frac{1}{\sqrt{2}}(\alpha + \beta).$$

Thus:

$$P = \left| \frac{\alpha + \beta}{\sqrt{2}} \right|^2 = \frac{1}{2}|\alpha + \beta|^2.$$

Chapter 12

Quantum Observables and Commutation Relations

Quantum Observables as Operators

In quantum mechanics, observables are not mere numbers but operators on a Hilbert space $\mathcal{H}$. These operators, typically represented by $\hat{A}$, $\hat{B}$, and others, correspond to measurable quantities such as position, momentum, and energy. A key attribute of these operators is their self-adjoint nature, defined by:

$$\hat{A} = \hat{A}^\dagger,$$

where $\hat{A}^\dagger$ denotes the adjoint of $\hat{A}$. The self-adjointness ensures that the eigenvalues of $\hat{A}$, which represent possible measurement outcomes, are real-valued.

Eigenvalues and Eigenvectors of Observables

For an observable $\hat{A}$, the equation

$$\hat{A}a = \lambda a$$

describes the eigenvalue λ and eigenvector a. These elements indicate that a measurement of the observable will yield λ with certainty if the system is initially in the state a.

Commutation Relations

Commutation relations reveal important information about the compatibility of simultaneous measurements. The commutator of two operators $\hat{A}$ and $\hat{B}$ is given by:

$$[\hat{A}, \hat{B}] = \hat{A}\hat{B} - \hat{B}\hat{A}.$$

If $[\hat{A}, \hat{B}] = 0$, the operators $\hat{A}$ and $\hat{B}$ are said to commute, implying that these observables can be measured simultaneously with arbitrary precision.

Canonical Commutation Relations

An essential example of commutation relations in quantum mechanics is the canonical commutation relation between position and momentum operators, $\hat{x}$ and $\hat{p}$:

$$[\hat{x}, \hat{p}] = i\hbar,$$

where $\hbar$ is the reduced Planck constant. This non-zero commutation relation signifies the fundamental quantum mechanical limitation on simultaneously knowing both position and momentum precisely.

The Uncertainty Principle

The uncertainty principle, formalized by Werner Heisenberg, is directly related to the commutation relations. For two non-commuting observables $\hat{A}$ and $\hat{B}$, the uncertainty principle is expressed as:

$$\sigma_A \sigma_B \geq \frac{1}{2} |\langle [\hat{A}, \hat{B}] \rangle|,$$

where σ_A and σ_B denote the standard deviations of the observables in a given state, and $\langle \cdot \rangle$ indicates the expectation value. This principle establishes a lower bound for the product of the uncertainties of two observables.

Expectation Values and Variance

The expectation value $\langle \hat{A} \rangle$ of an observable $\hat{A}$ in the state ψ is:

$$\langle \hat{A} \rangle = \langle \psi | \hat{A} | \psi \rangle.$$

The variance, a measure of uncertainty, is:

$$\text{Var}(\hat{A}) = \langle \hat{A}^2 \rangle - \langle \hat{A} \rangle^2.$$

Variance aids in quantifying the spread of an operator's outcomes around the mean.

Example: Position and Momentum

Consider a particle moving along one dimension with position operator $\hat{x}$ and momentum operator $\hat{p}$. The commutation relation:

$$[\hat{x}, \hat{p}] = i\hbar$$

indicates an intrinsic quantum limit on the precision of simultaneous measurements of position and momentum, as outlined by the Heisenberg uncertainty principle.
This leads to:

$$\sigma_x \sigma_p \geq \frac{\hbar}{2},$$

which imposes a limit on the product of position and momentum uncertainties.

Operator Algebra and Quantum Dynamics

Understanding the commutation relations is fundamental for exploring quantum dynamics. The time evolution of an operator $\hat{A}$ in the Heisenberg picture is given by:

$$\frac{d\hat{A}}{dt} = \frac{i}{\hbar}[\hat{H}, \hat{A}] + \frac{\partial \hat{A}}{\partial t},$$

where $\hat{H}$ is the Hamiltonian of the system. The commutation relations thus not only reveal measurement compatibility but also play a critical role in describing how quantum systems evolve over time.

Practice Problems 1

1. Show that the observable $\hat{A}$ is self-adjoint if $\hat{A} = \hat{A}^\dagger$, and explain why self-adjointness ensures real eigenvalues.

2. Prove that if two operators $\hat{A}$ and $\hat{B}$ commute, i.e. $[\hat{A}, \hat{B}] = 0$, then they share a common set of eigenvectors.

3. Derive the expression for the commutator of two operators $\hat{A}$ and $\hat{B}$: $[\hat{A}, \hat{B}]$, and discuss its implications in the context of measurements.

4. Utilize the canonical commutation relation $[\hat{x}, \hat{p}] = i\hbar$ to demonstrate the position-momentum uncertainty principle $\sigma_x \sigma_p \geq \frac{\hbar}{2}$.

5. Calculate the expectation value and variance for the observable $\hat{A}$ in the state ψ given by $\hat{A} = \begin{bmatrix} 2 & 0 \\ 0 & 3 \end{bmatrix}$,

and $\psi = \begin{bmatrix} 1/\sqrt{2} \\ 1/\sqrt{2} \end{bmatrix}$.

6. Explain the significance of the operator algebra in the Heisenberg picture and how it relates to the time evolution of quantum systems.

Answers 1

1. **Solution:** To show that $\hat{A}$ is self-adjoint, begin by showing:
$$\hat{A} = \hat{A}^\dagger$$
This means $(\hat{A}\psi)$ is identical to $\psi\hat{A}$, for all ψ. Self-adjoint operators are fundamental because they ensure the eigenvalues of $\hat{A}$ are real:

$$\hat{A}a = \lambda a \implies a\hat{A}a = \lambda aa$$

Taking the complex conjugate gives $\lambda^* = \lambda$, proving λ is real.

2. **Solution:** If $\hat{A}$ and $\hat{B}$ commute, $[\hat{A}, \hat{B}] = 0$. Suppose $\hat{A}a = \lambda a$, then:
$$\hat{A}\hat{B}a = \hat{B}\hat{A}a = \lambda\hat{B}a$$

Hence, $\hat{B}a$ is an eigenvector of $\hat{A}$ with eigenvalue λ. $\hat{A}$ and $\hat{B}$ share eigenvectors.

3. **Solution:** The commutator is:
$$[\hat{A}, \hat{B}] = \hat{A}\hat{B} - \hat{B}\hat{A}$$

This conveys uncertainty in joint measurements. If $[\hat{A}, \hat{B}] \neq 0$, the product of uncertainties is constrained by the uncertainty principle.

4. **Solution:** Use:
$$[\hat{x}, \hat{p}] = i\hbar$$
The uncertainty principle states:
$$\sigma_x \sigma_p \geq \frac{1}{2}|\langle[\hat{x}, \hat{p}]\rangle|$$
Calculating gives $|\langle i\hbar\rangle| = \hbar$, inserting:
$$\sigma_x \sigma_p \geq \frac{\hbar}{2}$$

5. **Solution:**

$$\langle \hat{A} \rangle = \langle \psi | \hat{A} | \psi \rangle = \begin{bmatrix} 1/\sqrt{2} & 1/\sqrt{2} \end{bmatrix} \begin{bmatrix} 2 & 0 \\ 0 & 3 \end{bmatrix} \begin{bmatrix} 1/\sqrt{2} \\ 1/\sqrt{2} \end{bmatrix}$$

$$= \begin{bmatrix} 1/\sqrt{2} & 1/\sqrt{2} \end{bmatrix} \begin{bmatrix} 1 \\ 3/2 \end{bmatrix}$$

$$= \frac{5}{2}$$

The variance is:

$$\text{Var}(\hat{A}) = \langle \hat{A}^2 \rangle - \langle \hat{A} \rangle^2 = \frac{13}{4} - \left(\frac{5}{2} \right)^2$$

$$= \frac{13}{4} - \frac{25}{4} = -3$$

Therefore:

$$\text{Var}(\hat{A}) = \frac{3}{4}.$$

6. **Solution:** In the Heisenberg picture, operators evolve by:

$$\frac{d\hat{A}}{dt} = \frac{i}{\hbar}[\hat{H}, \hat{A}] + \frac{\partial \hat{A}}{\partial t}$$

This shows the role of commutators in dynamics and observables' evolution.

Practice Problems 2

1. Consider the self-adjoint operator $\hat{A}$ representing an observable. Show that the expectation value $\langle \hat{A} \rangle$ is real for any state ψ.

2. Given the commutation relation $[\hat{x}, \hat{p}] = i\hbar$, derive the expression for the uncertainty principle relating the standard deviations σ_x and σ_p.

3. Prove that if two operators commute, $[\hat{A}, \hat{B}] = 0$, then they can be simultaneously diagonalized.

4. For a Hamiltonian $\hat{H}$, express the time evolution of an observable $\hat{A}$ in the Heisenberg picture and discuss its implications.

5. Demonstrate the eigenvalue equation $\hat{A}a = \lambda a$ implies that λ must be real if $\hat{A}$ is self-adjoint.

6. Consider the operators $\hat{x}$ and $\hat{p}$ satisfying $[\hat{x}, \hat{p}] = i\hbar$. Verify the canonical commutation relation by computing $[\hat{x}, \hat{p}]$ using their action on a wave function $\psi(x)$.

Answers 2

1. Consider the self-adjoint operator $\hat{A}$ representing an observable. Show that the expectation value $\langle \hat{A} \rangle$ is real for any state ψ. **Solution:** The expectation value of $\hat{A}$ is given by:

$$\langle \hat{A} \rangle = \langle \psi | \hat{A} | \psi \rangle$$

Since $\hat{A}$ is self-adjoint, $\hat{A} = \hat{A}^\dagger$, hence:

$$\langle \psi | \hat{A} | \psi \rangle = (\langle \psi | \hat{A} | \psi \rangle)^* = \langle \psi | \hat{A}^\dagger | \psi \rangle$$

This implies $\langle \hat{A} \rangle = \langle \hat{A} \rangle^*$, indicating that $\langle \hat{A} \rangle$ is real.

2. Given the commutation relation $[\hat{x}, \hat{p}] = i\hbar$, derive the expression for the uncertainty principle relating the standard deviations σ_x and σ_p. **Solution:** The uncertainty principle is derived using the relation:

$$\sigma_x \sigma_p \geq \frac{1}{2} |\langle [\hat{x}, \hat{p}] \rangle|$$

Substituting $[\hat{x}, \hat{p}] = i\hbar$, we get:

$$\sigma_x \sigma_p \geq \frac{1}{2} |\langle i\hbar \rangle| = \frac{\hbar}{2}$$

3. Prove that if two operators commute, $[\hat{A}, \hat{B}] = 0$, then they can be simultaneously diagonalized. **Solution:** If $[\hat{A}, \hat{B}] = 0$, then $\hat{A}$ and $\hat{B}$ share a common set of eigenvectors. Thus, for an eigenvector a of $\hat{A}$:

$$\hat{B}\hat{A}a = \hat{A}\hat{B}a$$

Since $\hat{A}a = \lambda a$ then:

$$\hat{B}\lambda a = \lambda \hat{B}a$$

meaning $\hat{B}a$ is also an eigenvector of $\hat{A}$. Hence $\hat{A}$ and $\hat{B}$ can be simultaneously diagonalized.

4. For a Hamiltonian $\hat{H}$, express the time evolution of an observable $\hat{A}$ in the Heisenberg picture and discuss its implications. **Solution:** In the Heisenberg picture, the time evolution of an observable $\hat{A}$ is:

$$\frac{d\hat{A}}{dt} = \frac{i}{\hbar}[\hat{H}, \hat{A}] + \frac{\partial \hat{A}}{\partial t}$$

This shows how $\hat{A}$ changes in time due to interactions described by $\hat{H}$. If $\hat{A}$ commutes with $\hat{H}$, its expectation value remains constant over time.

5. Demonstrate the eigenvalue equation $\hat{A}a = \lambda a$ implies that λ must be real if $\hat{A}$ is self-adjoint. **Solution:** Given $\hat{A}a = \lambda a$, taking the inner product with a:

$$\langle a | \hat{A} | a \rangle = \lambda \langle a | a \rangle$$

Since $\hat{A}$ is self-adjoint:

$$\langle a | \hat{A} | a \rangle = \langle a | \hat{A}^\dagger | a \rangle = \lambda \langle a | a \rangle^*$$

This implies $\lambda = \lambda^*$, so λ is real.

6. Consider the operators $\hat{x}$ and $\hat{p}$ satisfying $[\hat{x}, \hat{p}] = i\hbar$. Verify the canonical commutation relation by computing $[\hat{x}, \hat{p}]$ using their action on a wave function $\psi(x)$. **Solution:** Let $\hat{x}$ act as multiplication and $\hat{p} = -i\hbar \frac{d}{dx}$:

$$\hat{x}\hat{p}\psi(x) = \hat{x}\left(-i\hbar \frac{d\psi}{dx} \right) = -i\hbar x \frac{d\psi}{dx}$$

$$\hat{p}\hat{x}\psi(x) = -i\hbar \frac{d}{dx}(x\psi) = -i\hbar \left(\psi + x \frac{d\psi}{dx} \right)$$

So:

$$[\hat{x}, \hat{p}]\psi = -i\hbar x \frac{d\psi}{dx} + i\hbar(\psi + x \frac{d\psi}{dx}) = i\hbar\psi$$

Thus, $[\hat{x}, \hat{p}] = i\hbar$.

Practice Problems 3

1. Show that for a self-adjoint operator $\hat{A}$, the eigenvalues are real. Given that $\hat{A}a = \lambda a$, demonstrate that λ is real.

2. Given two operators $\hat{A}$ and $\hat{B}$, verify whether the following operators commute:

$$\hat{A} = \hat{x}, \quad \hat{B} = \hat{p}$$

where $\hat{x}$ and $\hat{p}$ are position and momentum operators respectively.

3. Prove the Heisenberg uncertainty principle for position $\hat{x}$ and momentum $\hat{p}$:

$$\sigma_x \sigma_p \geq \frac{\hbar}{2}$$

4. Calculate the expectation value of the position operator $\hat{x}$ in the state $\psi = \frac{1}{\sqrt{2}}(1 + 2)$ where 1 and 2 are orthonormal states.

5. Given $\hat{H} = \frac{\hat{p}^2}{2m} + V(\hat{x})$, find the time evolution of the position operator $\hat{x}$ in the Heisenberg picture.

6. Determine the variance of the observable $\hat{A}$ for a quantum state ϕ where $\hat{A}\phi = a\phi$ and a is a constant.

Answers 3

1. Show that for a self-adjoint operator $\hat{A}$, the eigenvalues are real. Given that $\hat{A}a = \lambda a$, demonstrate that λ is real.

   Solution: Consider the inner product of $\hat{A}a$ with a:

   $$a\hat{A}a = a\lambda a = \lambda aa$$

   Because $\hat{A}$ is self-adjoint, we also have:

   $$a\hat{A}a = a\hat{A}^\dagger a = (a\hat{A}a)^* \Rightarrow \lambda = \lambda^*$$

   Thus, λ is real.

2. Given two operators $\hat{A} = \hat{x}$ and $\hat{B} = \hat{p}$, verify whether they commute.

   Solution:
   $$[\hat{x}, \hat{p}] = \hat{x}\hat{p} - \hat{p}\hat{x}$$

   Using the position and momentum commutator relation:

   $$[\hat{x}, \hat{p}] = i\hbar \neq 0$$

   Hence, $\hat{x}$ and $\hat{p}$ do not commute.

3. Prove the Heisenberg uncertainty principle for position $\hat{x}$ and momentum $\hat{p}$:

   $$\sigma_x \sigma_p \geq \frac{\hbar}{2}$$

   Solution: Consider the operators $\hat{A} = \hat{x}$ and $\hat{B} = \hat{p}$:

   $$\sigma_x \sigma_p \geq \frac{1}{2}|\langle [\hat{A}, \hat{B}] \rangle|$$

$$\sigma_x \sigma_p \geq \frac{1}{2}|\langle i\hbar \rangle| = \frac{\hbar}{2}$$

Thus, the uncertainty relation is verified.

4. Calculate the expectation value of the position operator $\hat{x}$ in the state $\psi = \frac{1}{\sqrt{2}}(1 + 2)$.

   Solution:

   $$\langle \hat{x} \rangle = \psi \hat{x} \psi = \left(\frac{1}{\sqrt{2}}(1+2) \right) \hat{x} \left(\frac{1}{\sqrt{2}}(1+2) \right)$$

   $$= \frac{1}{2}(1\hat{x}1 + 1\hat{x}2 + 2\hat{x}1 + 2\hat{x}2)$$

   Assign values for each term and simplify to calculate.

5. Given $\hat{H} = \frac{\hat{p}^2}{2m} + V(\hat{x})$, find the time evolution of the position operator $\hat{x}$ in the Heisenberg picture.

   Solution:

   $$\frac{d\hat{x}}{dt} = \frac{i}{\hbar}[\hat{H}, \hat{x}]$$

   $$= \frac{i}{\hbar}\left[\frac{\hat{p}^2}{2m}, \hat{x} \right] + \frac{i}{\hbar}[V(\hat{x}), \hat{x}]$$

   The potential term commutes with $\hat{x}$:

   $$\Rightarrow \frac{d\hat{x}}{dt} = \frac{i}{\hbar}\frac{1}{2m}\left(\hat{p}^2\hat{x} - \hat{x}\hat{p}^2 \right)$$

   Simplifying using commutators yields the velocity as $\hat{v} = \frac{\hat{p}}{m}$.

6. Determine the variance of the observable $\hat{A}$ for a quantum state ϕ where $\hat{A}\phi = a\phi$ and a is a constant.

   Solution:

   $$\mathtt{Var}(\hat{A}) = \langle \hat{A}^2 \rangle - \langle \hat{A} \rangle^2$$

   Since $\hat{A}\phi = a\phi$,

   $$\langle \hat{A}^2 \rangle = a^2, \quad \langle \hat{A} \rangle = a$$

   Thus,

   $$\mathtt{Var}(\hat{A}) = a^2 - a^2 = 0$$

   Therefore, the variance is zero indicating no spread in measurements.

Chapter 13

Quantum Systems with Infinite Degrees of Freedom

Introduction to Infinite Degrees of Freedom

In quantum mechanics, physical systems are often characterized by a finite number of degrees of freedom. However, when extending to fields, as in quantum field theory (QFT), the systems exhibit infinite degrees of freedom. This shift necessitates conceptual and mathematical extensions beyond traditional quantum mechanics.

The Harmonic Oscillator as a Model

The quantum harmonic oscillator with finite degrees of freedom serves as a foundational model. The Hamiltonian is given by

$$\hat{H} = \frac{\hat{p}^2}{2m} + \frac{1}{2}m\omega^2\hat{x}^2,$$

where $\hat{x}$ is the position operator, $\hat{p}$ is the momentum operator, m is the mass, and ω is the angular frequency. The solution of this system relies on solving the Schrödinger equation

$$\hat{H}\psi = E\psi.$$

Field Theory: Introducing Continuity

To model fields, consider classical fields defined at every point in space. Each point contributes a degree of freedom. Transitioning to quantum fields involves extending the principles of quantum mechanics to field operators $\hat{\phi}(x)$ instead of discrete variables.

Quantum Fields

A quantum field $\hat{\phi}(x,t)$ is an operator-valued distribution over spacetime. The field's dynamics can be described by a Lagrangian density $\mathcal{L}(\phi, \partial_\mu \phi)$, leading to field equations via the Euler-Lagrange equation:

$$\frac{\partial}{\partial x^\mu}\left(\frac{\partial \mathcal{L}}{\partial(\partial_\mu \phi)}\right) - \frac{\partial \mathcal{L}}{\partial \phi} = 0.$$

Canonical Quantization

Canonical quantization involves promoting fields to operators and quantifying via canonical commutation relations. For a field $\hat{\phi}(x)$, the field momentum $\hat{\pi}(x)$ is:

$$\hat{\pi}(x) = \frac{\partial \mathcal{L}}{\partial(\partial_t \phi(x))}.$$

The fundamental commutation relations are then:

$$[\hat{\phi}(x), \hat{\pi}(x')] = i\hbar\delta(x - x'),$$

where $\delta(x - x')$ is the Dirac delta function.

Basis for Quantum Field Theory

Transitioning from quantum mechanics to QFT alters the foundational structure representing states and observables. The vacuum state 0 and creation $\hat{a}^\dagger(k)$ and annihilation $\hat{a}(k)$ operators form the basis, satisfying:

$$[\hat{a}(k), \hat{a}^\dagger(k')] = \delta(k - k').$$

These operators construct Fock space by creating and annihilating particle excitations.

Spectral Theory in Infinite Dimensional Spaces

Spectral theory is crucial for analyzing states in infinite-dimensional spaces. Operators are either bounded or unbounded, with the spectral theorem providing a way to decompose self-adjoint operators. For an operator $\hat{A}$, the spectral decomposition is:

$$\hat{A} = \int \lambda \, dE_\lambda,$$

where E_λ are projection operators.

Mathematical Challenges

The mathematical description of systems with infinite degrees of freedom, especially concerning convergence and operator domains, presents challenges. This necessitates a rigorous treatment where Hilbert space theory and distributional approaches, such as constructing *rigged Hilbert spaces*, are employed.

Transition to Quantum Field Theory

Transitioning to QFT requires incorporating classical fields' dynamics with quantum mechanics' probabilistic nature. Quantum fields replace position and momentum operators, enabling the formulation of interactions at the field level. Path integrals offer a reformulation through functional integration over field configurations, crucial for non-perturbative methods.

Data: Initialize classical field configurations
Result: Calculate path integral over field configurations
Discretize spacetime into a lattice;
Approximate the action $\mathcal{S}[\phi]$ from $\mathcal{L}(\phi, \partial_\mu \phi)$;
Integrate: $\int \mathcal{D}\phi \, e^{\frac{i}{\hbar}\mathcal{S}[\phi]}$;
Select configurations minimizing action for field equations solutions;

Practice Problems 1

1. Describe how canonical quantization is applied to fields to obtain quantum field operators. What role do commutation relations play?

2. Explain the process of constructing the Fock space from creation and annihilation operators. How does this relate to quantum states?

3. Derive the expression for the field momentum $\hat{\pi}(x)$ from the Lagrangian density $\mathcal{L}$. Why is it essential in canonical quantization?

4. Discuss the significance of the spectral decomposition theorem in the context of quantum field theory (QFT). How does it facilitate the analysis of quantum states?

5. Demonstrate how path integrals are used in the context of QFT to calculate observables. What is the significance of discretizing spacetime?

6. Explain the concept of infinite degrees of freedom in quantum fields. How does this change the mathematical techniques used compared to finite-dimensional quantum systems?

Answers 1

1. **Canonical Quantization of Fields:**
   Solution: Canonical quantization involves promoting classical fields to operators, using commutation relations to encapsulate quantum behavior. For a field $\hat{\phi}(x)$, quantization requires the field to become an operator and the imposition of canonical commutation relations such as:

   $$[\hat{\phi}(x), \hat{\pi}(x')] = i\hbar\delta(x - x').$$

   These relations ensure that the algebra of operators reflects the quantum nature of phenomena, allowing for the calculation of observables using these operators in a quantum setting.

2. **Construction of Fock Space:**
   Solution: The Fock space is constructed by utilizing creation $\hat{a}^\dagger(k)$ and annihilation $\hat{a}(k)$ operators. These operators add or remove particle states at momentum k:

   $$[\hat{a}(k), \hat{a}^\dagger(k')] = \delta(k - k').$$

   Starting from the vacuum state 0, repeated application of the creation operator generates higher particle states. This hierarchical structure represents quantum states with variable particle numbers, essential for quantum systems involving many particles.

3. **Expression for Field Momentum:**
   Solution: Given the Lagrangian density $\mathcal{L}(\phi, \partial_\mu\phi)$, the field momentum $\hat{\pi}(x)$ is defined as:

   $$\hat{\pi}(x) = \frac{\partial\mathcal{L}}{\partial(\partial_t\phi(x))}.$$

   This expression is derived using the Euler-Lagrange formalism and is crucial in canonical quantization because $\hat{\pi}(x)$ becomes the conjugate momentum to the field $\hat{\phi}(x)$, entering the commutation relations and ensuring consistent quantum dynamics.

4. **Spectral Decomposition Theorem in QFT:**
   Solution: The spectral decomposition theorem allows the analysis of self-adjoint operators, such as Hamiltonians, in terms of their spectral measures:

   $$\hat{A} = \int \lambda \, dE_\lambda.$$

   This decomposition helps in understanding the spectrum of eigenvalues representing physically observable quantities. In QFT, it assists in analyzing continuous spectra common in field theories and enables expressions for state evolutions and measurable properties.

5. **Path Integrals in QFT:**
   Solution: Path integrals rephrase quantum mechanics using functional integration:

   $$\int \mathcal{D}\phi \, e^{\frac{i}{\hbar}S[\phi]}.$$

   Discretizing spacetime into a lattice simplifies calculations, allowing numerical approximations for complex integrals. This method provides insights into non-perturbative regimes, beyond the reach of traditional perturbative techniques.

6. **Infinite Degrees of Freedom:**
   Solution: Quantum fields, unlike quantum particles, possess infinite degrees of freedom as they are defined at every point in space. This necessitates advanced mathematical tools like functional analysis and distributions. Concepts like rigged Hilbert spaces and operator theory are employed to manage the complexities arising from this infinite dimensionality, highlighting the need for rigor in field quantization.

Practice Problems 2

1. Consider the Hamiltonian of a quantum harmonic oscillator:

   $$\hat{H} = \frac{\hat{p}^2}{2m} + \frac{1}{2}m\omega^2\hat{x}^2.$$

   Derive the commutation relation between the position operator $\hat{x}$ and the Hamiltonian $\hat{H}$.

2. Given the definition of the field momentum $\hat{\pi}(x) = \frac{\partial \mathcal{L}}{\partial(\partial_t \phi(x))}$, show how canonical quantization leads to the commutation relation $[\hat{\phi}(x), \hat{\pi}(x')] = i\hbar\delta(x - x')$.

3. Using the Lagrangian density $\mathcal{L}(\phi, \partial_\mu \phi) = \frac{1}{2}(\partial^\mu \phi \partial_\mu \phi - m^2\phi^2)$, derive the Euler-Lagrange equation for the field ϕ.

4. Compute the spectral decomposition of a self-adjoint operator $\hat{A}$ using the spectral theorem. For simplicity, assume $\hat{A}$ has a discrete spectrum.

5. Given the creation and annihilation operators $\hat{a}^\dagger(k)$ and $\hat{a}(k)$ satisfying $[\hat{a}(k), \hat{a}^\dagger(k')] = \delta(k-k')$, derive the expression for the number operator $\hat{N}$.

6. For the quantum field $\hat{\phi}(x,t)$, discuss the transformation to the Fock space representation by explaining the role of the vacuum state 0.

Answers 2

1. The commutation relation between the position operator $\hat{x}$ and the Hamiltonian $\hat{H}$ can be derived using:

$$[\hat{x}, \hat{H}] = [\hat{x}, \frac{\hat{p}^2}{2m} + \frac{1}{2}m\omega^2\hat{x}^2].$$

Since $[\hat{x}, \hat{x}^2] = 0$, we focus on:

$$[\hat{x}, \hat{p}^2] = \hat{x}\hat{p}^2 - \hat{p}^2\hat{x}.$$

Using $[\hat{x}, \hat{p}] = i\hbar$, we find:

$$[\hat{x}, \hat{p}^2] = \hat{x}\hat{p}\hat{p} - \hat{p}\hat{x}\hat{p} = \hat{x}\hat{p}^2 - \hat{p}(\hat{p}\hat{x} + i\hbar) = i\hbar\hat{p}.$$

Therefore,

$$[\hat{x}, \hat{H}] = \frac{i\hbar}{2m}\hat{p},$$

confirming that $[\hat{x}, \hat{H}]$ describes a velocity term in quantum mechanics.

2. By definition of the field momentum,

$$\hat{\pi}(x) = \frac{\partial \mathcal{L}}{\partial(\partial_t \phi(x))}.$$

For the Lagrangian density $\mathcal{L} = \frac{1}{2}(\partial^\mu \phi \partial_\mu \phi - m^2 \phi^2)$,

$$\hat{\pi}(x) = \partial^0 \phi = \dot{\phi}.$$

Canonical quantization requires:

$$[\hat{\phi}(x), \hat{\pi}(x')] = [\hat{\phi}(x), \dot{\phi}(x')] = i\hbar\delta(x - x').$$

This commutation follows from the requirement that fields and field momenta obey canonical commutation relations similar to the Heisenberg uncertainty relations.

3. The Euler-Lagrange equation is derived by:

$$\frac{\partial}{\partial x^\mu}\left(\frac{\partial \mathcal{L}}{\partial(\partial_\mu \phi)}\right) - \frac{\partial \mathcal{L}}{\partial \phi} = 0.$$

For $\mathcal{L}(\phi, \partial_\mu \phi) = \frac{1}{2}(\partial^\mu \phi \partial_\mu \phi - m^2 \phi^2)$,

$$\frac{\partial \mathcal{L}}{\partial(\partial_\mu \phi)} = \partial^\mu \phi.$$

Thus,

$$\partial_\mu \partial^\mu \phi + m^2 \phi = 0,$$

resulting in the Klein-Gordon equation.

4. The spectral theorem states:

$$\hat{A} = \sum_\lambda \lambda E_\lambda,$$

where E_λ are projection operators. For each eigenvalue λ, the operator can be expressed in terms of its spectral decomposition using projectors built from its eigenfunctions.

5. For the number operator $\hat{N}$,

$$\hat{N} = \hat{a}^\dagger(k)\hat{a}(k).$$

Using the commutation relations,

$$[\hat{a}(k), \hat{a}^\dagger(k')] = \delta(k - k'),$$

one writes:

$$\hat{N} = \hat{a}^\dagger(k)\hat{a}(k).$$

This operator counts the number of quanta or excitations in the mode k.

6. The vacuum state 0 plays a fundamental role in Fock space. It is defined by:

$$\hat{a}(k)0 = 0, \quad \forall k.$$

The creation operator $\hat{a}^\dagger(k)$ acts on the vacuum to create one-particle states:

$$1_k = \hat{a}^\dagger(k)0.$$

The Fock space is constructed by repeated applications of creation operators, forming a complete basis for quantum states in field theory.

Practice Problems 3

1. Describe how the canonical quantization process applies to a scalar field $\phi(x)$ in $1+1$ dimensions, starting from the classical Lagrangian density:

$$\mathcal{L} = \frac{1}{2}(\partial_t \phi)^2 - \frac{1}{2}(\partial_x \phi)^2 - \frac{1}{2}m^2\phi^2.$$

2. Given the quantum harmonic oscillator Hamiltonian

$$\hat{H} = \frac{\hat{p}^2}{2m} + \frac{1}{2}m\omega^2\hat{x}^2,$$

   demonstrate the quantization of the system via the introduction of ladder operators $\hat{a}$ and $\hat{a}^\dagger$.

3. Prove that the commutation relation $[\hat{a}, \hat{a}^\dagger] = 1$ holds for the creation and annihilation operators in the context of the harmonic oscillator.

4. Show how to derive the field operator decomposition for a free scalar field $\phi(x)$ in terms of creation and annihilation operators over all modes.

5. Describe how the spectral theorem applies to bounded operators in the context of quantum mechanics, specifically for an operator $\hat{A}$ on a Hilbert space.

6. Explain the role of the Dirac delta function $\delta(x - x')$ within the canonical commutation relations for field operators.

Answers 3

1. Describe how the canonical quantization process applies to a scalar field $\phi(x)$ in $1 + 1$ dimensions.
   Solution: Begin with the classical Lagrangian density:

$$\mathcal{L} = \frac{1}{2}(\partial_t \phi)^2 - \frac{1}{2}(\partial_x \phi)^2 - \frac{1}{2}m^2\phi^2$$

The conjugate momentum is:

$$\pi(x) = \frac{\partial \mathcal{L}}{\partial(\partial_t \phi)} = \partial_t \phi$$

The field and momentum are promoted to operators $\hat{\phi}(x)$ and $\hat{\pi}(x)$, with commutation relations:

$$[\hat{\phi}(x), \hat{\pi}(x')] = i\hbar\delta(x - x')$$

Canonical quantization thus introduces non-commuting field operators, foundational for QFT.

2. Given the quantum harmonic oscillator Hamiltonian, demonstrate the quantization via ladder operators. **Solution:** Start with the Hamiltonian:

$$\hat{H} = \frac{\hat{p}^2}{2m} + \frac{1}{2}m\omega^2\hat{x}^2$$

Define $\hat{a}$ and $\hat{a}^\dagger$ as:

$$\hat{a} = \sqrt{\frac{m\omega}{2\hbar}}\left(\hat{x} + \frac{i}{m\omega}\hat{p}\right), \quad \hat{a}^\dagger = \sqrt{\frac{m\omega}{2\hbar}}\left(\hat{x} - \frac{i}{m\omega}\hat{p}\right)$$

Recast $\hat{x}$ and $\hat{p}$:

$$\hat{x} = \sqrt{\frac{\hbar}{2m\omega}}(\hat{a} + \hat{a}^\dagger), \quad \hat{p} = -i\sqrt{\frac{\hbar m\omega}{2}}(\hat{a} - \hat{a}^\dagger)$$

Substituting gives $\hat{H} = \hbar\omega(\hat{a}^\dagger\hat{a} + \frac{1}{2})$.

3. Prove the commutation relation $[\hat{a}, \hat{a}^\dagger] = 1$. **Solution:** Using definitions:

$$[\hat{a}, \hat{a}^\dagger] = \left[\sqrt{\frac{m\omega}{2\hbar}} \left(\hat{x} + \frac{i}{m\omega} \hat{p} \right), \sqrt{\frac{m\omega}{2\hbar}} \left(\hat{x} - \frac{i}{m\omega} \hat{p} \right) \right]$$

Substitute $[\hat{x}, \hat{p}] = i\hbar$ and simplify:

$$[\hat{a}, \hat{a}^\dagger] = 1$$

4. Derive the field operator decomposition for a scalar field $\phi(x)$. **Solution:** Express $\phi(x)$ in terms of modes:

$$\hat{\phi}(x,t) = \int \frac{dk}{\sqrt{2\pi}} \frac{1}{\sqrt{2\omega_k}} (\hat{a}(k) e^{i(kx - \omega_k t)} + \hat{a}^\dagger(k) e^{-i(kx - \omega_k t)})$$

Here, $\omega_k = \sqrt{k^2 + m^2}$.

5. Describe how the spectral theorem applies to bounded operators in quantum mechanics. **Solution:** For a bounded self-adjoint operator $\hat{A}$:

$$\hat{A} = \int \lambda \, dE_\lambda$$

where E_λ is a projection-valued measure, decomposing $\hat{A}$ into its eigenvalues.

6. Explain the role of $\delta(x - x')$ in canonical commutation relations. **Solution:** The Dirac delta function $\delta(x - x')$ ensures locality:

$$[\hat{\phi}(x), \hat{\pi}(x')] = i\hbar \delta(x - x')$$

It dictates that field operators at different points commute like independent degrees of freedom.

Chapter 14

Canonical Quantization of Fields

Classical Field Theory

Classical fields are typically described by a Lagrangian density $\mathcal{L}(\phi, \partial_\mu \phi) = \frac{1}{2}(\partial^\mu \phi \partial_\mu \phi - m^2 \phi^2)$ for a scalar field $\phi(x)$. The field equations are derived via the Euler-Lagrange equation:

$$\frac{\partial}{\partial x^\mu}\left(\frac{\partial \mathcal{L}}{\partial(\partial_\mu \phi)}\right) - \frac{\partial \mathcal{L}}{\partial \phi} = 0.$$

Canonical Quantization Process

To quantize a field, promote the classical field $\phi(x)$ to an operator $\hat{\phi}(x)$. The conjugate momentum is introduced:

$$\hat{\pi}(x) = \frac{\partial \mathcal{L}}{\partial(\partial_t \phi(x))}.$$

The process of canonical quantization imposes commutation relations on these operators:

$$[\hat{\phi}(x), \hat{\pi}(x')] = i\hbar \delta(x - x').$$

Field Operators and Their Commutation Relations

In quantum field theory, field operators are used to describe fields at each point x in space:

$$\hat{\phi}(x) = \int \frac{dk}{\sqrt{2\pi}} \frac{1}{\sqrt{2\omega_k}}(\hat{a}(k)e^{ikx} + \hat{a}^\dagger(k)e^{-ikx}).$$

The canonical commutation relations play a crucial role:

$$[\hat{a}(k), \hat{a}^\dagger(k')] = \delta(k - k').$$

Creation and Annihilation Operators

Creation $(\hat{a}^\dagger(k))$ and annihilation $(\hat{a}(k))$ operators alter the particle number in a given mode k. The Fock space is constructed using these operators. Starting from the vacuum state 0 defined by:

$$\hat{a}(k)0 = 0,$$

particles are created by:

$$n = \frac{1}{\sqrt{n!}}(\hat{a}^\dagger(k))^n 0.$$

Quantization Example: Real Scalar Field

Consider a real scalar field with the Lagrangian:

$$\mathcal{L} = \frac{1}{2}(\partial_\mu \phi \partial^\mu \phi - m^2 \phi^2).$$

The conjugate momentum is:

$$\pi(x) = \frac{\partial \mathcal{L}}{\partial(\partial_t \phi)} = \partial_t \phi.$$

The Hamiltonian density is given by:

$$\mathcal{H} = \pi \partial_t \phi - \mathcal{L}.$$

Promotion to operators with $\hat{\phi}(t, \mathbf{x})$ and $\hat{\pi}(t, \mathbf{x})$ yields the commutation relations:

$$[\hat{\phi}(t, \mathbf{x}), \hat{\pi}(t, \mathbf{y})] = i\hbar \delta(\mathbf{x} - \mathbf{y}).$$

Role of Commutation Relations

Commutation relations ensure the consistency of quantum mechanics in quantum field theory. For field operators, they are reminiscent of the Heisenberg uncertainty principle, establishing a fundamental limit to the precision with which pairs of complementary variables (like field and field momentum) can be known.

Formulation in Fock Space

The Fock space formalism assembles all possible states of a quantum field. Creation and annihilation operators act to increase or decrease particle numbers. The quantum state is represented as a superposition of Fock states.

Algorithm 3: Constructing Fock Space States

Data: Initial vacuum state 0
Result: Fock space with all particle numbers
foreach *momentum k* **do**
  $\lfloor$ Apply $\hat{a}^\dagger(k)$ to 0 repeatedly to generate states

---

Practice Problems 1

1. Derive the classical field equation from the given Lagrangian density:

$$\mathcal{L}(\phi, \partial_\mu \phi) = \frac{1}{2}(\partial^\mu \phi \partial_\mu \phi - m^2 \phi^2)$$

2. Demonstrate the steps to promote a classical field to a quantum field operator:

3. Show the computation of the conjugate momentum for the scalar field:

$$\mathcal{L} = \frac{1}{2}(\partial_\mu \phi \partial^\mu \phi - m^2 \phi^2)$$

4. Verify the canonical commutation relations for field operators:

$$[\hat{\phi}(x), \hat{\pi}(x')] = i\hbar\delta(x - x')$$

5. Calculate the Hamiltonian density for the given Lagrangian:

$$\mathcal{L} = \frac{1}{2}(\partial_\mu \phi \partial^\mu \phi - m^2 \phi^2)$$

6. Illustrate the action of creation and annihilation operators on the vacuum state:

Answers 1

1. Derive the classical field equation from the given Lagrangian density:

   Solution: Start with the Lagrangian density:

   $$\mathcal{L}(\phi, \partial_\mu \phi) = \frac{1}{2}(\partial^\mu \phi \partial_\mu \phi - m^2 \phi^2).$$

   The Euler-Lagrange equation is:

   $$\frac{\partial}{\partial x^\mu}\left(\frac{\partial \mathcal{L}}{\partial(\partial_\mu \phi)}\right) - \frac{\partial \mathcal{L}}{\partial \phi} = 0.$$

   Calculate $\frac{\partial \mathcal{L}}{\partial(\partial_\mu \phi)}$:

   $$\frac{\partial \mathcal{L}}{\partial(\partial_\mu \phi)} = \partial^\mu \phi.$$

   Then:

   $$\frac{\partial}{\partial x^\mu}(\partial^\mu \phi) = \partial_\mu \partial^\mu \phi.$$

   Compute $\frac{\partial \mathcal{L}}{\partial \phi}$:

   $$\frac{\partial \mathcal{L}}{\partial \phi} = -m^2 \phi.$$

   Substitute into the Euler-Lagrange equation:

   $$\partial_\mu \partial^\mu \phi + m^2 \phi = 0.$$

   This is the Klein-Gordon equation.

2. Demonstrate the steps to promote a classical field to a quantum field operator:

   Solution: Begin with the classical field $\phi(x)$. Promote $\phi(x)$ to an operator $\hat{\phi}(x)$. Classical field equations transform to operator equations. Example: Classical $\phi(x)$ updates to $\hat{\phi}(x)$.

3. Show the computation of the conjugate momentum for the scalar field:

   Solution: Start with the Lagrangian:

   $$\mathcal{L} = \frac{1}{2}(\partial_\mu \phi \partial^\mu \phi - m^2 \phi^2).$$

   Define conjugate momentum $\pi(x)$:

   $$\pi(x) = \frac{\partial \mathcal{L}}{\partial(\partial_t \phi(x))}.$$

   Calculate:

   $$\pi(x) = \frac{1}{2} \cdot 2\partial_t \phi = \partial_t \phi.$$

4. Verify the canonical commutation relations for field operators:

   Solution: Utilize the relations:

   $$[\hat{\phi}(x), \hat{\pi}(x')] = i\hbar\delta(x - x').$$

   Confirm through computation:

   $$[\hat{\phi}(x), \hat{\pi}(x')] = i\hbar\delta(x - x').$$

   Consistency is checked via basic commutation proofs.

5. Calculate the Hamiltonian density for the given Lagrangian:

   Solution: Start with the Lagrangian:

   $$\mathcal{L} = \frac{1}{2}(\partial_\mu\phi\partial^\mu\phi - m^2\phi^2).$$

   Define Hamiltonian density $\mathcal{H}$:

   $$\mathcal{H} = \pi\partial_t\phi - \mathcal{L}.$$

   Substitute:

   $$\mathcal{H} = \pi^2 + \frac{1}{2}m^2\phi^2 = \frac{1}{2}(\partial_t\phi)^2 + \frac{1}{2}m^2\phi^2.$$

6. Illustrate the action of creation and annihilation operators on the vacuum state:

   Solution: Vacuum state:

   $$0.$$

   Annihilation operators:

   $$\hat{a}(k)0 = 0.$$

   Creation operators act:

   $$n = \frac{1}{\sqrt{n!}}(\hat{a}^\dagger(k))^n 0.$$

   Each application of $\hat{a}^\dagger(k)$ increases particle number.

Practice Problems 2

1. Derive the field equation for a scalar field $\phi(x)$ using the Lagrangian density:

   $$\mathcal{L} = \frac{1}{2}(\partial_\mu\phi\partial^\mu\phi - m^2\phi^2).$$

2. Calculate the commutation relation for the field operator $\hat{\phi}(x)$ and its conjugate momentum $\hat{\pi}(x')$.

3. Express the field operator $\hat{\phi}(x)$ in terms of creation and annihilation operators for a real scalar field.

4. Show that the Hamiltonian density for the given scalar field Lagrangian is:

$$\mathcal{H} = \frac{1}{2}(\pi^2 + (\nabla\phi)^2 + m^2\phi^2).$$

5. Prove that the Fock space is constructed from the vacuum state using creation operators $\hat{a}^\dagger(k)$.

6. Verify the normalization condition for a single-particle state in Fock space: $\langle 1_k | 1_{k'} \rangle = \delta(k - k')$.

Answers 2

1. Derive the field equation for a scalar field $\phi(x)$ using the Lagrangian density:

$$\mathcal{L} = \frac{1}{2}(\partial_\mu \phi \partial^\mu \phi - m^2\phi^2).$$

Solution:
To derive the field equation, apply the Euler-Lagrange equation:

$$\frac{\partial}{\partial x^\mu}\left(\frac{\partial \mathcal{L}}{\partial(\partial_\mu \phi)}\right) - \frac{\partial \mathcal{L}}{\partial \phi} = 0.$$

Compute $\frac{\partial \mathcal{L}}{\partial(\partial_\mu \phi)} = \partial^\mu \phi$.
Differentiate: $\frac{\partial}{\partial x^\mu}(\partial^\mu \phi) = \Box \phi$.
Compute $\frac{\partial \mathcal{L}}{\partial \phi} = -m^2 \phi$.
Thus, $\Box \phi + m^2 \phi = 0$, which is the Klein-Gordon equation.

2. Calculate the commutation relation for the field operator $\hat{\phi}(x)$ and its conjugate momentum $\hat{\pi}(x')$.
   Solution:
   The commutation relation is given by:

$$[\hat{\phi}(x), \hat{\pi}(x')] = i\hbar \delta(x - x').$$

   Substitute $\hat{\pi}(x) = \partial_t \hat{\phi}(x)$ in the canonical quantization framework.
   Verifying this requires the definition and properties of the Dirac delta function.

3. Express the field operator $\hat{\phi}(x)$ in terms of creation and annihilation operators for a real scalar field.
   Solution:
   Express $\hat{\phi}(x)$ as:

$$\hat{\phi}(x) = \int \frac{dk}{\sqrt{2\pi}} \frac{1}{\sqrt{2\omega_k}}(\hat{a}(k)e^{ikx} + \hat{a}^\dagger(k)e^{-ikx}).$$

   This involves the Fourier transform of field operators and normalizing factors based on energy, ω_k.

4. Show that the Hamiltonian density for the given scalar field Lagrangian is:

$$\mathcal{H} = \frac{1}{2}(\pi^2 + (\nabla\phi)^2 + m^2\phi^2).$$

   Solution:
   Start with the Lagrangian: $\mathcal{L} = \frac{1}{2}(\partial_\mu \phi \partial^\mu \phi - m^2 \phi^2)$.

$$\pi = \partial_t \phi,$$

$$\mathcal{H} = \pi \partial_t \phi - \mathcal{L}.$$

   Substitute and simplify:

$$= \frac{1}{2}(\pi^2 + (\nabla\phi)^2 + m^2\phi^2).$$

5. Prove that the Fock space is constructed from the vacuum state using creation operators $\hat{a}^\dagger(k)$.
   Solution:
   Begin with the vacuum state 0 where:
$$\hat{a}(k)0 = 0.$$

   Apply $\hat{a}^\dagger(k)$ repeatedly:

$$n = \frac{1}{\sqrt{n!}}(\hat{a}^\dagger(k))^n 0.$$

   These states n form a complete basis for the Fock space.

6. Verify the normalization condition for a single-particle state in Fock space: $\langle 1_k | 1_{k'} \rangle = \delta(k - k')$.
   Solution:
   Compute the one-particle state:
$$1_k = \hat{a}^\dagger(k)0.$$

Calculate the inner product:
$$\langle 1_k | 1_{k'} \rangle = \langle 0 | \hat{a}(k) \hat{a}^\dagger(k') | 0 \rangle.$$

Use the commutation relation:
$$[\hat{a}(k), \hat{a}^\dagger(k')] = \delta(k - k'),$$

yielding:
$$\langle 1_k | 1_{k'} \rangle = \delta(k - k').$$

Practice Problems 3

1. Derive the Euler-Lagrange equation for a scalar field from the given Lagrangian density:
$$\mathcal{L}(\phi, \partial_\mu \phi) = \frac{1}{2} \left(\partial^\mu \phi \partial_\mu \phi - m^2 \phi^2 \right).$$

2. Given the commutation relations for field operators, derive the expression for the canonical commutation relation:
$$[\hat{\phi}(x), \hat{\pi}(x')] = i\hbar \delta(x - x').$$

3. Explain the significance of the creation and annihilation operators in the context of quantum fields and describe their algebraic properties.

4. Construct the quantum Hamiltonian for a real scalar field starting from the Lagrangian density:

$$\mathcal{L} = \frac{1}{2}(\partial_\mu \phi \partial^\mu \phi - m^2 \phi^2).$$

5. Discuss the importance of the Fock space in quantum field theory and outline the process of creating multi-particle states from the vacuum state.

6. Examine the role of the commutation relations in ensuring the consistency of the canonical quantization approach in quantum field theory.

Answers 3

1. Derive the Euler-Lagrange equation for a scalar field from the given Lagrangian density:

$$\mathcal{L}(\phi, \partial_\mu \phi) = \frac{1}{2} \left(\partial^\mu \phi \partial_\mu \phi - m^2 \phi^2 \right).$$

Solution: The Euler-Lagrange equation for a field is expressed as:

$$\frac{\partial}{\partial x^\mu} \left(\frac{\partial \mathcal{L}}{\partial(\partial_\mu \phi)} \right) - \frac{\partial \mathcal{L}}{\partial \phi} = 0.$$

Calculate $\frac{\partial \mathcal{L}}{\partial(\partial_\mu \phi)}$:

$$\frac{\partial \mathcal{L}}{\partial(\partial_\mu \phi)} = \partial^\mu \phi.$$

Substituting into the Euler-Lagrange equation yields:

$$\frac{\partial}{\partial x^\mu}(\partial^\mu \phi) - (-m^2\phi) = 0,$$

which simplifies to the Klein-Gordon equation:

$$\Box\phi + m^2\phi = 0.$$

2. Given the commutation relations for field operators, derive the expression for the canonical commutation relation:

$$[\hat{\phi}(x), \hat{\pi}(x')] = i\hbar\delta(x - x').$$

   Solution: The field $\hat{\phi}(x)$ and its conjugate momentum $\hat{\pi}(x)$ form a canonical pair. According to quantum mechanics, their commutator is given by:

$$[\hat{\phi}(x), \hat{\pi}(x')] = \hat{\phi}(x)\hat{\pi}(x') - \hat{\pi}(x')\hat{\phi}(x).$$

   In the canonical quantization framework, imposing:

$$[\hat{\phi}(x), \hat{\pi}(x')] = i\hbar\delta(x - x'),$$

   ensures that the field and its momentum satisfy a relation analogous to the Heisenberg uncertainty principle. This relation is essential for quantizing classical fields into operators.

3. Explain the significance of the creation and annihilation operators in the context of quantum fields and describe their algebraic properties. **Solution:** Creation ($\hat{a}^\dagger(k)$) and annihilation ($\hat{a}(k)$) operators are key elements in quantum field theory, enabling the quantization of fields over all space:
   - $\hat{a}^\dagger(k)$ adds a quantum of excitation to the state at momentum k.
   - $\hat{a}(k)$ removes a quantum of excitation from the state at momentum k.

   Their algebraic properties follow the canonical commutation relations:

$$[\hat{a}(k), \hat{a}^\dagger(k')] = \delta(k - k').$$

   These operators permit the mathematical construction of a Fock space, which describes many-particle states and elegantly captures the probabilistic nature of quantum mechanics.

4. Construct the quantum Hamiltonian for a real scalar field starting from the Lagrangian density:

$$\mathcal{L} = \frac{1}{2}(\partial_\mu\phi\partial^\mu\phi - m^2\phi^2).$$

   Solution: The Hamiltonian is obtained by the Legendre transform:

$$\mathcal{H} = \pi\dot{\phi} - \mathcal{L}.$$

   The conjugate momentum π is:

$$\pi = \frac{\partial\mathcal{L}}{\partial(\partial_t\phi)} = \partial_t\phi.$$

   Substitute π into the Hamiltonian:

$$\mathcal{H} = \frac{1}{2}(\partial_t\phi)^2 + \frac{1}{2}(\nabla\phi)^2 + \frac{1}{2}m^2\phi^2.$$

   The Hamiltonian density represents total energy, capturing both kinetic ($\partial_t\phi$) and potential ($\nabla\phi, m^2\phi^2$) energy contributions.

5. Discuss the importance of the Fock space in quantum field theory and outline the process of creating multi-particle states from the vacuum state. **Solution:** The Fock space represents the Hilbert space of all possible states in quantum field theory:

- It accommodates states with varying particle numbers, unlike traditional quantum mechanics confined to a single-particle description.
- It builds upon the vacuum state 0, accessing an infinite set by acting with creation operators.

The process:
- Begin with the vacuum state: 0.
- Apply creation operators: $\hat{a}^\dagger(k)0$ for each momentum k.
- Multi-particle states n_k result from multiple applications, e.g., $\frac{1}{\sqrt{n!}}(\hat{a}^\dagger(k))^n 0$.

6. Examine the role of the commutation relations in ensuring the consistency of the canonical quantization approach in quantum field theory. **Solution:** Commutation relations are crucial for maintaining the integrity of quantum mechanical predictions through field quantization.

- They mirror the foundational commutation relations in quantum mechanics between observables and momenta (e.g., $[x, p] = i\hbar$).
- The relations $[\hat{\phi}(x), \hat{\pi}(y)] = i\hbar\delta(x - y)$ guarantee the non-commutative structure essential for the uncertainty principle, ensuring consistent measurements and evolving states according to quantum mechanics.
- They sustain the calculation of observable properties such as energy and momentum to obey expected physical laws.

Chapter 15

Boson Fields in Hilbert Space Framework

Introduction to Bosonic Fields

Bosonic fields, fundamental in the study of quantum field theory, are characterized by integer spin. The quantization of such fields, described by appropriate Lagrangian densities, leads to the introduction of creation and annihilation operators in a Hilbert space framework. This approach facilitates the encoding of particle interactions and properties.

Quantization of Scalar Boson Fields

Consider a real scalar field $\phi(x)$ described by the Lagrangian density:

$$\mathcal{L} = \frac{1}{2} \left(\partial_\mu \phi \partial^\mu \phi - m^2 \phi^2 \right).$$

The quantization process involves promoting the classical field $\phi(x)$ to an operator $\hat{\phi}(x)$ within a Hilbert space. The corresponding conjugate momentum operator is:

$$\hat{\pi}(x) = \frac{\partial \mathcal{L}}{\partial(\partial_t \phi)} = \partial_t \hat{\phi}(x).$$

Commutation relations are established as:

$$[\hat{\phi}(t, \mathbf{x}), \hat{\pi}(t, \mathbf{y})] = i\hbar \delta(\mathbf{x} - \mathbf{y}).$$

These relations ensure adherence to quantum mechanical principles.

Creation and Annihilation Operators

The field operator $\hat{\phi}(x)$ can be represented using creation $\hat{a}^\dagger(k)$ and annihilation $\hat{a}(k)$ operators as follows:

$$\hat{\phi}(x) = \int \frac{d^3 k}{\sqrt{(2\pi)^3 2\omega_k}} \left(\hat{a}(k) e^{ik \cdot x} + \hat{a}^\dagger(k) e^{-ik \cdot x} \right),$$

where $\omega_k = \sqrt{\mathbf{k}^2 + m^2}$. These operators satisfy the canonical commutation relations:

$$[\hat{a}(k), \hat{a}^\dagger(k')] = \delta^3(\mathbf{k} - \mathbf{k}'), \quad [\hat{a}(k), \hat{a}(k')] = [\hat{a}^\dagger(k), \hat{a}^\dagger(k')] = 0.$$

Fock Space Construction

Fock space provides a convenient framework to describe multi-particle states. The vacuum state 0 is defined such that:

$$\hat{a}(k)0 = 0 \quad \text{for all } k.$$

A single-particle state is obtained by applying a creation operator to the vacuum:

$$1_k = \hat{a}^\dagger(k)0.$$

Multi-particle states are constructed through successive applications of creation operators:

$$n_{k_1}, n_{k_2}, \ldots = \frac{1}{\sqrt{n_1! \, n_2! \ldots}} (\hat{a}^\dagger(k_1))^{n_1} (\hat{a}^\dagger(k_2))^{n_2} \ldots 0.$$

Fock space is characterized by the direct sum of these n-particle subspaces.

Connection to Quantum Fields: Scalar and Vector Fields

The quantization framework readily applies to vector fields. For a vector field $\mathbf{A}(x)$, a similar approach yields:

$$\hat{\mathbf{A}}(x) = \sum_{\lambda=1}^{3} \int \frac{d^3k}{\sqrt{(2\pi)^3 2\omega_k}} \left(\hat{a}_\lambda(k)\epsilon_\lambda e^{ik\cdot x} + \hat{a}_\lambda^\dagger(k)\epsilon_\lambda^* e^{-ik\cdot x} \right),$$

where ϵ_λ are polarization vectors and $\hat{a}_\lambda(k)$ obey analogous commutation relations.

Applications in Scalar and Vector Fields

The formalism of Fock space and operator quantization is pivotal in describing interactions in scalar and vector fields. The understanding of elementary particles and forces involves these quantized fields. Applications include calculating scattering amplitudes and understanding particle creation and annihilation processes within collider experiments. The Hilbert space framework offers a robust suite of mathematical tools to model these quantum phenomena.

Practice Problems 1

1. Derive the expression for the commutation relation between the field operator $\hat{\phi}(x)$ and its conjugate momentum $\hat{\pi}(y)$ in the quantization of a scalar boson field.

2. Demonstrate how the creation operator $\hat{a}^\dagger(k)$ acts on a Fock space vacuum state 0 to yield a one-particle state 1_k.

3. Show that the integral representation of the field operator $\hat{\phi}(x)$ satisfies the Klein-Gordon equation for free fields.

4. Explain the significance of the commutation relations among the creation and annihilation operators in preserving bosonic field statistics.

5. Derive the form of the Hamiltonian operator for a quantized real scalar field using creation and annihilation operators.

6. Calculate the norm of a two-particle state $1_{k_1}, 1_{k_2}$ in Fock space and discuss its physical interpretation.

Answers 1

1. **Derive the expression for the commutation relation between the field operator $\hat{\phi}(x)$ and its conjugate momentum $\hat{\pi}(y)$:**

   Solution: The commutation relation is given by:

$$[\hat{\phi}(t, \mathbf{x}), \hat{\pi}(t, \mathbf{y})] = i\hbar\delta(\mathbf{x} - \mathbf{y}).$$

To derive this, we note that $\hat{\pi}(t, \mathbf{y}) = \partial_t\hat{\phi}(t, \mathbf{y})$. The canonical quantization ensures that:

$$[\hat{\phi}(t, \mathbf{x}), \partial_t\hat{\phi}(t, \mathbf{y})] = i\hbar\delta(\mathbf{x} - \mathbf{y}),$$

reflecting the localized interaction property of field quanta.

2. **Demonstrate how the creation operator $\hat{a}^\dagger(k)$ acts on a Fock space vacuum state 0:**
   Solution: The creation operator $\hat{a}^\dagger(k)$ adds a quantum of momentum k to the vacuum state:

$$\hat{a}^\dagger(k)0 = 1_k.$$

   This procedure generates a single-particle state with definite momentum k, which is a basic construction in quantum field theory to represent particles.

3. **Show that the integral representation of the field operator $\hat{\phi}(x)$ satisfies the Klein-Gordon equation:**
   Solution: The Klein-Gordon equation is:

$$(\partial^\mu\partial_\mu + m^2)\phi(x) = 0.$$

   With the field operator given by:

$$\hat{\phi}(x) = \int \frac{d^3k}{\sqrt{(2\pi)^3 2\omega_k}} \left(\hat{a}(k)e^{ik\cdot x} + \hat{a}^\dagger(k)e^{-ik\cdot x}\right),$$

   substituting this into the Klein-Gordon equation results in a dispersion relation:

$$\omega_k^2 = \mathbf{k}^2 + m^2,$$

   hence confirming that the operator satisfies the equation for free fields.

4. **Explain the significance of the commutation relations among creation and annihilation operators:**
   Solution: The commutation relations:

$$[\hat{a}(k), \hat{a}^\dagger(k')] = \delta^3(\mathbf{k} - \mathbf{k}'),$$

   ensure that the statistics of bosonic fields are preserved. Specifically, these relations imply that bosons can occupy the same quantum state, reflecting the indistinguishable and symmetric nature of bosons, fundamental to quantum statistics.

5. **Derive the form of the Hamiltonian operator for a quantized real scalar field:**
   Solution: Starting from:

$$\mathcal{H} = \frac{1}{2}(\hat{\pi}^2 + (\nabla\hat{\phi})^2 + m^2\hat{\phi}^2),$$

   substituting in terms of $\hat{a}(k)$ and $\hat{a}^\dagger(k)$:

$$\hat{H} = \int d^3k\,\omega_k\left(\hat{a}^\dagger(k)\hat{a}(k) + \frac{1}{2}\right),$$

   confirming that each quantum of ω_k corresponds to an energy particle created or annihilated by the operators.

6. **Calculate the norm of a two-particle state $1_{k_1}, 1_{k_2}$ in Fock space:**

   Solution: The norm of the state:

   $$1_{k_1}, 1_{k_2} = \hat{a}^\dagger(k_1)\hat{a}^\dagger(k_2)0,$$

   is given by:

   $$1_{k_1}, 1_{k_2}|1_{k_1}, 1_{k_2} = 1,$$

   indicating normalization in Fock space, ensuring a consistent representation of identical particles without further scaling factors.

Practice Problems 2

1. Show that the commutation relations for the creation and annihilation operators imply the uncertainty principle. Specifically, consider the operators $\hat{a}(k)$ and $\hat{a}^\dagger(k)$ and demonstrate the connection to the uncertainty principle. 4cm

2. Derive the expression for the number operator $\hat{N} = \hat{a}^\dagger\hat{a}$ in the context of a Fock space and verify that its eigenvalues correspond to the number of particles. 4cm

3. Prove that the vacuum state 0 is normalized in a Fock space and that $0\hat{a}^\dagger\hat{a}0 = 0$. 4cm

4. Show how creation and annihilation operators transform under a spatial translation, i.e., $\hat{\phi}(x + a)$ in terms of $\hat{\phi}(x)$. 4cm

5. Validate the completeness relation in Fock space for the basis vectors composed of multi-particle states. 4cm

6. Use the canonical commutation relations to demonstrate that the field operator $\hat{\phi}(x)$ satisfies the Klein-Gordon equation. 4cm

Answers 2

1. The commutation relations for the creation and annihilation operators read:

   $$[\hat{a}(k), \hat{a}^\dagger(k')] = \delta^3(\mathbf{k} - \mathbf{k}'), \quad [\hat{a}(k), \hat{a}(k')] = [\hat{a}^\dagger(k), \hat{a}^\dagger(k')] = 0.$$

   Solution: For operators $\hat{X}$ and $\hat{P}$ with commutator $[\hat{X}, \hat{P}] = i\hbar$, we have the uncertainty relation $\Delta X \Delta P \geq \frac{\hbar}{2}$. Identify $\hat{X} \equiv \frac{\hat{a}+\hat{a}^\dagger}{\sqrt{2}}$ and $\hat{P} \equiv i\frac{\hat{a}-\hat{a}^\dagger}{\sqrt{2}}$. The canonical commutator can be rewritten to match the form $[\hat{X}, \hat{P}] = i\hbar$, confirming the uncertainty principle.

2. The number operator is defined as:
   $$\hat{N} = \hat{a}^\dagger\hat{a}.$$

   Solution: Calculate the action on a state n:

   $$\hat{N}n = \hat{a}^\dagger\hat{a}n = nn,$$

   implying that n is an eigenstate with eigenvalue n. This shows the number operator counts the number of particles.

152

3. For the vacuum state normalization,
$$0|0 = 1,$$
and
$$0\hat{a}^\dagger\hat{a}0 = 000 = 0.$$

Solution: The operator $\hat{a}$ annihilates the vacuum:
$$\hat{a}0 = 0.$$

For the expectation value,
$$0\hat{a}^\dagger\hat{a}0 = 0\hat{a}\hat{a}^\dagger0 - 00 = 0.$$

4. For the translation operator $e^{-i\hat{P}a}$,
$$\hat{\phi}(x + a) = e^{-i\hat{P}a}\hat{\phi}(x)e^{i\hat{P}a}.$$

Solution: In Fourier space:
$$\hat{\phi}(x + a) = \int \frac{d^3k}{\sqrt{(2\pi)^3 2\omega_k}}\left(\hat{a}(k)e^{ik\cdot x}e^{ika} + \hat{a}^\dagger(k)e^{-ik\cdot x}e^{-ika}\right).$$

5. Completeness relation:
$$\sum_n nn = \mathbb{I}.$$

Solution: Rudimentary resolution of identity:
$$\sum_{n=0}^{\infty} \frac{(\hat{a}^\dagger)^n}{\sqrt{n!}}00\frac{\hat{a}^n}{\sqrt{n!}} = \mathbb{I}.$$

6. Using the Klein-Gordon equation $(\Box + m^2)\hat{\phi}(x) = 0$, **Solution:** Differentiate
$$\Box\hat{\phi}(x) = \int \frac{d^3k}{\sqrt{(2\pi)^3 2\omega_k}}\left(-k^2\hat{a}(k)e^{ik\cdot x} + k^2\hat{a}^\dagger(k)e^{-ik\cdot x}\right),$$

which simplifies using energy relation $\omega_k^2 = \mathbf{k}^2 + m^2$.

Practice Problems 3

1. Verify the commutation relation for the field operator and its conjugate momentum operator:
$$[\hat{\phi}(t, \mathbf{x}), \hat{\pi}(t, \mathbf{y})] = i\hbar\delta(\mathbf{x} - \mathbf{y}).$$

2. Show how a field operator $\hat{\phi}(x)$ can be expressed in terms of creation and annihilation operators:

$$\hat{\phi}(x) = \int \frac{d^3k}{\sqrt{(2\pi)^3 2\omega_k}} \left(\hat{a}(k)e^{ik\cdot x} + \hat{a}^\dagger(k)e^{-ik\cdot x} \right).$$

3. Derive the canonical commutation relations for creation and annihilation operators:

$$[\hat{a}(k), \hat{a}^\dagger(k')] = \delta^3(\mathbf{k} - \mathbf{k}'), \quad [\hat{a}(k), \hat{a}(k')] = [\hat{a}^\dagger(k), \hat{a}^\dagger(k')] = 0.$$

4. Construct a single-particle state using the creation operator:

$$1_k = \hat{a}^\dagger(k)0.$$

5. Explain how multi-particle states are formed in Fock space:

$$n_{k_1}, n_{k_2}, \ldots = \frac{1}{\sqrt{n_1! \, n_2! \, \ldots}} (\hat{a}^\dagger(k_1))^{n_1} (\hat{a}^\dagger(k_2))^{n_2} \ldots 0.$$

6. Demonstrate the quantization framework applied to a vector field $\mathbf{A}(x)$:

$$\hat{\mathbf{A}}(x) = \sum_{\lambda=1}^{3} \int \frac{d^3k}{\sqrt{(2\pi)^3 2\omega_k}} \left(\hat{a}_\lambda(k)\epsilon_\lambda e^{ik\cdot x} + \hat{a}_\lambda^\dagger(k)\epsilon_\lambda^* e^{-ik\cdot x} \right).$$

Answers 3

1. **Solution:** To verify the commutation relation for the field operator and its conjugate momentum operator, we start by writing:

$$\hat{\pi}(x) = \partial_t \hat{\phi}(x).$$

Using the canonical commutation relation:

$$[\hat{\phi}(t, \mathbf{x}), \hat{\pi}(t, \mathbf{y})] = \left[\hat{\phi}(t, \mathbf{x}), \partial_t \hat{\phi}(t, \mathbf{y}) \right] = i\hbar\delta(\mathbf{x} - \mathbf{y}).$$

This arises from their derivations from the field Lagrangian and is a fundamental aspect of quantum field theory.

2. **Solution:** The field operator $\hat{\phi}(x)$ is expressed in terms of creation and annihilation operators as follows:

$$\hat{\phi}(x) = \int \frac{d^3k}{\sqrt{(2\pi)^3 2\omega_k}} \left(\hat{a}(k)e^{ik\cdot x} + \hat{a}^\dagger(k)e^{-ik\cdot x} \right).$$

This expression decomposes the field into its Fourier components, where the operators $\hat{a}(k)$ and $\hat{a}^\dagger(k)$ account for the absorption and emission of particles.

3. **Solution:** The canonical commutation relations for creation and annihilation operators are derived from:

$$[\hat{a}(k), \hat{a}^\dagger(k')] = \delta^3(\mathbf{k} - \mathbf{k}').$$

These relations ensure that a particle is either added or removed from the specified momentum mode, maintaining consistency with the underlying quantum mechanics.

4. **Solution:** A single-particle state is constructed by applying the creation operator to the vacuum state:

$$1_k = \hat{a}^\dagger(k)0.$$

This operation elevates the vacuum to a state with one particle of momentum k, showcasing the creation aspect of the operator in Fock space.

5. **Solution:** Multi-particle states are constructed by:

$$n_{k_1}, n_{k_2}, \ldots = \frac{1}{\sqrt{n_1! n_2! \ldots}} (\hat{a}^\dagger(k_1))^{n_1} (\hat{a}^\dagger(k_2))^{n_2} \ldots 0.$$

By applying the creation operators successively, particles are introduced into the system, and normalization factors ensure state orthogonality.

6. **Solution:** The quantization framework applied to vector fields is shown as:

$$\hat{\mathbf{A}}(x) = \sum_{\lambda=1}^{3} \int \frac{d^3k}{\sqrt{(2\pi)^3 2\omega_k}} \left(\hat{a}_\lambda(k)\epsilon_\lambda e^{ik\cdot x} + \hat{a}_\lambda^\dagger(k)\epsilon_\lambda^* e^{-ik\cdot x} \right).$$

Here, ϵ_λ are the polarization vectors, signifying different polarizations or orientations for the vector fields, providing a comprehensive model for electromagnetic waves.

Chapter 16

Fermion Fields and Anticommutation Relations

Introduction to Fermionic Fields

Fermionic fields are integral components of quantum field theory, representing particles that obey Fermi-Dirac statistics and possess half-integer spin. The treatment of these fields diverges from bosonic fields due to their inherent antisymmetry, necessitating the employment of anticommutation relations instead of commutation relations.

Anticommutation Relations

Fermionic creation and annihilation operators, denoted by $\hat{b}^\dagger(k)$ and $\hat{b}(k)$ respectively, satisfy anticommutation relations given by:

$$\{\hat{b}(k), \hat{b}^\dagger(k')\} = \delta^3(\mathbf{k} - \mathbf{k}'), \quad \{\hat{b}(k), \hat{b}(k')\} = \{\hat{b}^\dagger(k), \hat{b}^\dagger(k')\} = 0.$$

These relations reflect the Pauli exclusion principle, asserting that no two fermions can occupy the same quantum state.

Quantization of Fermionic Fields

The quantization of fermionic fields involves promoting classical fields to operators that act within a Hilbert space. For a Dirac field $\psi(x)$, the field operator can be expressed as:

$$\hat{\psi}(x) = \int \frac{d^3k}{(2\pi)^{3/2}\sqrt{2E_k}} \sum_s \left[\hat{b}(k,s)u(k,s)e^{-ik\cdot x} + \hat{d}^\dagger(k,s)v(k,s)e^{ik\cdot x} \right],$$

where $u(k,s)$ and $v(k,s)$ are spinors corresponding to the particle and antiparticle solutions, and $E_k = \sqrt{\mathbf{k}^2 + m^2}$ is the energy.

Fermi-Dirac Fock Space

Fermi-Dirac Fock space is constructed to accommodate the antisymmetry of fermion states, ensuring the wavefunction changes sign upon exchange of any two particles. The vacuum state 0 adheres to the condition:

$$\hat{b}(k,s)0 = \hat{d}(k,s)0 = 0,$$

for all momentum k and spin s. Single-particle states are formed by applying a creation operator to the vacuum state:

$$1_{\mathbf{k},s} = \hat{b}^\dagger(\mathbf{k},s)0,$$

$$1_{\bar{\mathbf{k}},\bar{s}} = \hat{d}^\dagger(\mathbf{k},s)0.$$

Multi-particle states arise through further applications of creation operators, with the total antisymmetry maintained by the anticommutation properties.

Applications of Anticommutation Relations

The significance of anticommutation relations in fermionic systems is evident in various phenomena, such as electron behavior in atoms and the formation of band structures in solid state physics. These relations govern transitions and the overall stability of fermionic systems, underscoring their role in the broader quantum field theory framework.

The combination of anticommutation relations and quantized fields provides a robust mechanism to describe and predict fermionic behavior, essential for understanding and developing new quantum theories.

Practice Problems 1

1. Verify the anticommutation relation for fermionic creation and annihilation operators:

$$\{\hat{b}(k), \hat{b}^\dagger(k')\} = \hat{b}(k)\hat{b}^\dagger(k') + \hat{b}^\dagger(k')\hat{b}(k)$$

and confirm that it equals $\delta^3(\mathbf{k} - \mathbf{k}')$.

2. Show that the fermionic operators satisfy the Pauli exclusion principle by demonstrating:

$$\{\hat{b}(k), \hat{b}(k')\} = 0.$$

3. Derive the expression for the Dirac field operator $\hat{\psi}(x)$ given:

$$\hat{\psi}(x) = \int \frac{d^3k}{(2\pi)^{3/2}\sqrt{2E_k}} \sum_s \left[\hat{b}(k,s)u(k,s)e^{-ik\cdot x} + \hat{d}^\dagger(k,s)v(k,s)e^{ik\cdot x} \right].$$

4. Construct the vacuum state 0 in a Fermi-Dirac Fock space and verify:

$$\hat{b}(k,s)0 = 0 \quad \text{and} \quad \hat{d}(k,s)0 = 0.$$

5. Can a state in Fermi-Dirac Fock space be symmetric under exchange of two identical particles? Justify your answer.

6. Discuss the role of anticommutation relations in explaining electron pairing in superconductivity.

Answers 1

1. **Solution to Problem 1:** We want to compute the anticommutation:

$$\{\hat{b}(k), \hat{b}^\dagger(k')\} = \hat{b}(k)\hat{b}^\dagger(k') + \hat{b}^\dagger(k')\hat{b}(k)$$

By definition, the anticommutation relation is:

$$\hat{b}(k)\hat{b}^\dagger(k') = \delta^3(\mathbf{k} - \mathbf{k}') - \hat{b}^\dagger(k')\hat{b}(k)$$

Adding both sides, we get:

$$\hat{b}(k)\hat{b}^\dagger(k') + \hat{b}^\dagger(k')\hat{b}(k) = \delta^3(\mathbf{k} - \mathbf{k}')$$

Hence, the required relation is confirmed.

2. **Solution to Problem 2:** To show:

$$\{\hat{b}(k), \hat{b}(k')\} = \hat{b}(k)\hat{b}(k') + \hat{b}(k')\hat{b}(k) = 0$$

By the definition of anticommutation relations for fermions, no two identical fermions can occupy the same state:

$$\hat{b}(k)\hat{b}(k') = -\hat{b}(k')\hat{b}(k)$$

So:

$$\hat{b}(k)\hat{b}(k') + \hat{b}(k')\hat{b}(k) = 0$$

Thus demonstrating the Pauli exclusion principle.

3. **Solution to Problem 3:** The Dirac field operator $\hat{\psi}(x)$ is derived from the transformation of a classical Dirac field into operators:

$$\hat{\psi}(x) = \int \frac{d^3k}{(2\pi)^{3/2}\sqrt{2E_k}} \sum_s \left[\hat{b}(k,s)u(k,s)e^{-ik\cdot x} + \hat{d}^\dagger(k,s)v(k,s)e^{ik\cdot x} \right]$$

where $u(k,s)$ represents positive energy solutions and $v(k,s)$ negative energy solutions. The creation and annihilation operators reflect quantum states.

4. **Solution to Problem 4:** In a Fermi-Dirac Fock space, the vacuum state 0 satisfies:

$$\hat{b}(k,s)0 = \hat{d}(k,s)0 = 0$$

which implies no particles or antiparticles exist in the vacuum state, which is the foundation for building other Fock states through creation operators.

5. **Solution to Problem 5:** A state in Fermi-Dirac Fock space cannot be symmetric under the exchange of two identical fermions. By definition, the exchange of two identical fermions results in an antisymmetric state, consistent with:

$$\psi(x_1, x_2) = -\psi(x_2, x_1)$$

so the symmetry rule doesn't hold due to anticommutation.

6. **Solution to Problem 6:** Anticommutation relations are fundamental in explaining electron pairing in superconductivity via Cooper pairs. Paired electrons (Cooper pairs) form a Bose-Einstein condensate due to attractive interactions at low temperatures; the antisymmetric nature of these pairs underlines the microscopic theory of superconductivity.

Practice Problems 2

1. Verify that the anticommutation relations for the fermionic creation operator $\hat{b}^\dagger(k)$ and annihilation operator $\hat{b}(k)$ are consistent with the Pauli exclusion principle by demonstrating that:

$$\{\hat{b}^\dagger(k), \hat{b}^\dagger(k')\} = 0 \quad \text{and} \quad \{\hat{b}(k), \hat{b}(k')\} = 0$$

2. Derive the expression for the Dirac field operator $\hat{\psi}(x)$ in momentum space given the classical Dirac spinors u and v.

3. Show how the requirements of antisymmetry in multi-particle states lead to their representation in the Fermi-Dirac Fock space.

4. Consider the single-particle state $1_{\mathbf{k},s}$. Demonstrate how you verify this state follows the anticommutation relations by computing $\hat{b}^\dagger(\mathbf{k}, s)\hat{b}^\dagger(\mathbf{k}', s')0$.

5. Explain how the anticommutation relations contribute to the stability of electronic configurations in atoms.

6. Discuss the significance of anti-symmetry for fermions and how it affects the overall mathematical formulation within quantum field theory.

Answers 2

1. **Solution:** To verify that the anticommutation relations imply fermionic behavior, consider:

$$\{\hat{b}^\dagger(k), \hat{b}^\dagger(k')\} = \hat{b}^\dagger(k)\hat{b}^\dagger(k') + \hat{b}^\dagger(k')\hat{b}^\dagger(k) = 0$$

Since $\hat{b}^\dagger(k)$ and $\hat{b}^\dagger(k')$ are operators that create fermions, their antisymmetry ensures two fermions cannot occupy the same state, reflecting the Pauli exclusion principle.

2. **Solution:** Starting from classical fields, we quantize them as:

$$\hat{\psi}(x) = \int \frac{d^3k}{(2\pi)^{3/2}\sqrt{2E_k}} \sum_s \left[\hat{b}(k,s)u(k,s)e^{-ik\cdot x} + \hat{d}^\dagger(k,s)v(k,s)e^{ik\cdot x} \right]$$

Each term reflects positive and negative frequency solutions using creation and annihilation operators over momentum k and spin s.

3. **Solution:** The construction of Fermi-Dirac Fock space involves ensuring antisymmetric wave functions. Consider:

$$\psi_{\text{multi}} = \hat{b}^\dagger(k_1)\hat{b}^\dagger(k_2)\cdots\hat{b}^\dagger(k_N)0$$

Swapping any two $\hat{b}^\dagger$ operators yields a sign change due to anticommutation, ensuring the total wave function remains antisymmetric.

4. **Solution:** For verifying the state properties under anticommutation:

$$\hat{b}^\dagger(\mathbf{k},s)\hat{b}^\dagger(\mathbf{k}',s')0 = -\hat{b}^\dagger(\mathbf{k}',s')\hat{b}^\dagger(\mathbf{k},s)0$$

As $\{\hat{b}^\dagger(\mathbf{k},s), \hat{b}^\dagger(\mathbf{k}',s')\} = 0$, it confirms no two fermions can share identical quantum numbers, adhering to Pauli exclusion principles.

5. **Solution:** Anticommutation relations ensure stability by preventing electrons from occupying the same state, restricting electrons within atomic orbitals and establishing specific configurations leading to observed chemical properties.

6. **Solution:** Antisymmetry underpins the definitions within fermionic fields. Every fermionic system adheres to the antisymmetric property as required by the algebra of operators, enforcing physical laws like Pauli exclusion and influencing the theoretical structure of the quantum field systems.

Practice Problems 3

1. Construct the anticommutation relations for the fermionic creation $\hat{b}^\dagger(k)$ and annihilation $\hat{b}(k)$ operators and verify if they satisfy the Pauli exclusion principle condition.

2. Derive the expression for the Dirac field operator $\hat{\psi}(x)$ and identify the role of $\hat{b}(k, s)$ and $\hat{d}^\dagger(k, s)$ in this context.

3. Explain the construction of a Fermi-Dirac Fock space and describe the method to form a single-particle state.

4. Demonstrate how the antisymmetry in fermionic states is maintained by showing the effect of exchanging two particles in a multi-particle Fock state.

5. Discuss the implications of anticommutation relations in solid-state physics, specifically relating to electron behavior in atoms.

6. Show how the anticommutation properties contribute to ensuring the stability of fermionic systems within quantum field theory.

Answers 3

1. Construct the anticommutation relations for the fermionic creation $\hat{b}^\dagger(k)$ and annihilation $\hat{b}(k)$ operators and verify if they satisfy the Pauli exclusion principle condition.

   Solution: By definition, the anticommutation relations for fermionic operators are:

   $$\{\hat{b}(k), \hat{b}^\dagger(k')\} = \delta^3(\mathbf{k} - \mathbf{k}'),$$
   $$\{\hat{b}(k), \hat{b}(k')\} = \{\hat{b}^\dagger(k), \hat{b}^\dagger(k')\} = 0.$$

   The first relation ensures that the creation and annihilation operators for fermions satisfy the Pauli exclusion principle by indicating that no two fermions can occupy the same state. This antisymmetry is the key feature that differentiates fermions from bosons.

2. Derive the expression for the Dirac field operator $\hat{\psi}(x)$ and identify the role of $\hat{b}(k,s)$ and $\hat{d}^\dagger(k,s)$ in this context.

   Solution: The Dirac field operator is given by:

   $$\hat{\psi}(x) = \int \frac{d^3k}{(2\pi)^{3/2}\sqrt{2E_k}} \sum_s \left[\hat{b}(k,s)u(k,s)e^{-ik\cdot x} + \hat{d}^\dagger(k,s)v(k,s)e^{ik\cdot x} \right],$$

   Here, $\hat{b}(k,s)$ is the annihilation operator for a fermion with momentum k and spin s, while $\hat{d}^\dagger(k,s)$ is the creation operator for an antifermion (hole) with the same quantum numbers. The $u(k,s)$ and $v(k,s)$ are the respective Dirac spinors.

3. Explain the construction of a Fermi-Dirac Fock space and describe the method to form a single-particle state.

   Solution: Fermi-Dirac Fock space is constructed by applying creation operators to a vacuum state 0, which is annihilated by all destruction operators:

$$\hat{b}(k,s)0 = \hat{d}(k,s)0 = 0.$$

A single-particle state is created by acting with a creation operator on the vacuum state:

$$1_{\mathbf{k},s} = \hat{b}^\dagger(\mathbf{k},s)0,$$

$$1_{\bar{\mathbf{k}},\bar{s}} = \hat{d}^\dagger(\mathbf{k},s)0.$$

4. Demonstrate how the antisymmetry in fermionic states is maintained by showing the effect of exchanging two particles in a multi-particle Fock state.

   Solution: Consider a two-body state $1,2 = 1_{\mathbf{k}_1,s_1} \otimes 1_{\mathbf{k}_2,s_2}$. When the fermionic operators are exchanged, this results in:

   $$2,1 = \hat{b}^\dagger(\mathbf{k}_2,s_2)\hat{b}^\dagger(\mathbf{k}_1,s_1)0 = -\hat{b}^\dagger(\mathbf{k}_1,s_1)\hat{b}^\dagger(\mathbf{k}_2,s_2)0 = -1,2,$$

   thus satisfying the antisymmetry requirement.

5. Discuss the implications of anticommutation relations in solid-state physics, specifically relating to electron behavior in atoms.

   Solution: In solid-state physics, anticommutation relations manifest in the Pauli exclusion principle, which is crucial for determining electron configurations in atoms and hence their chemical properties. The principle ensures that electrons fill discrete energy levels before occupying higher ones, leading to the formation of band structures in solids.

6. Show how the anticommutation properties contribute to ensuring the stability of fermionic systems within quantum field theory.

   Solution: Anticommutation properties prevent fermions from occupying the same quantum state, leading to repulsion at short distances and overall stability in systems such as electrons in atoms. This avoidance of overlap plays a fundamental role in determining the stability and behavior of matter, as well as contributing to the highly structured nature of fermionic systems within quantum field theory.

Chapter 17

Spectral Theory in Quantum Field Theory

Introduction to Spectral Theory

Spectral theory is the study of the spectrum of operators, especially self-adjoint operators, which play a crucial role in quantum mechanics and quantum field theory. Here, the focus lies on different types of spectra, including point, continuous, and residual spectra, in the context of quantum field operators.

Field Operators in Quantum Field Theory

In quantum field theory, field operators are used to describe quantum fields. These operators are usually infinite-dimensional and can be either bounded or unbounded, which affects their spectral properties.

Let $\hat{\phi}(x)$ be a scalar field operator. The time evolution of such a field is governed by a Hamiltonian $\hat{H}$, expressed as an integral over a spatial slice:

$$\hat{H} = \int d^3x \, \mathcal{H}(\hat{\phi}(x), \nabla\hat{\phi}(x), \hat{\pi}(x)),$$

where $\mathcal{H}$ is the Hamiltonian density and $\hat{\pi}(x)$ is the conjugate momentum operator.

Spectral Theorem for Self-Adjoint Operators

The spectral theorem provides a framework for understanding self-adjoint operators, central to quantum mechanics. For a self-adjoint linear operator $\hat{A}$, the spectral theorem states there exists a projection-valued measure $E(\lambda)$ such that

$$\hat{A} = \int_{\sigma(\hat{A})} \lambda \, dE(\lambda),$$

where $\sigma(\hat{A})$ is the spectrum of $\hat{A}$.

Energy Spectrum Analysis

In quantum field theory, analyzing the energy spectrum of field operators gives insights into particle states. The eigenvalues of the Hamiltonian operator $\hat{H}$ correspond to the possible energy levels of the system.

Consider a quantum field described by an operator $\hat{H}$ acting on a Hilbert space $\mathcal{H}$. The energy spectrum can often be decomposed into discrete and continuous parts:

$$\sigma(\hat{H}) = \sigma_d(\hat{H}) \cup \sigma_c(\hat{H}),$$

where $\sigma_d(\hat{H})$ represents discrete eigenvalues and $\sigma_c(\hat{H})$ represents the continuous spectrum.

Applications to Particle Physics

In particle physics, the spectral properties of quantum field operators provide essential insights into the nature of elementary particles and their interactions. The Hamiltonian's spectrum, in particular, reveals information about particle mass and the stability of states.

The application of spectral theory includes analyzing particle mass as eigenvalues in the spectrum of the field's Hamiltonian. The bound states correspond to discrete spectra and identified as particles. For instance, using quantum chromodynamics (QCD), the study of bound states like hadrons involves examining the spectral properties of the related Hamiltonian.

Implications for Quantum Field Theory

Spectral theory extends beyond basic applications, impacting renormalization and scattering theory in quantum field theory. By understanding the spectrum of operators, one can explore phenomena such as resonance and stability of quantum states, which are crucial for theoretical predictions in particle physics.

The spectral representation can be applied to propagator calculations aiding in the understanding of particle interactions at various energy levels.

Spectral Representation in Quantum Field Theory

The spectral representation is crucial for deriving propagators, which describe how quantum states evolve over spacetime. For a field operator propagator $D(x - y)$, the spectral representation is expressed as:

$$D(x - y) = \int_{-\infty}^{+\infty} \frac{d\mu(\sigma)}{2\pi} e^{-i\sigma(x-y)},$$

where $\mu(\sigma)$ is the spectral density function, providing insights into the distribution of states in the spectrum.

This representation illustrates how different modes of the field propagate and decay over spacetime, essential for predictions and computations in quantum field theory.

Practice Problems 1

1. Consider the self-adjoint operator $\hat{A}$ with spectrum $\sigma(\hat{A}) = \{1, 2, 3\}$. Write the spectral decomposition of $\hat{A}$ using projection operators.

2. Given a Hamiltonian $\hat{H}$ with discrete energy spectrum $\sigma_d(\hat{H}) = \{E_1, E_2, E_3\}$, describe how these energy eigenvalues are related to the physical states of the quantum field theory.

3. Demonstrate the application of the spectral representation to a scalar field propagator $D(x - y)$ by expressing it in terms of the spectral density function.

4. Prove that a self-adjoint operator $\hat{B}$ has a real spectrum, briefly explaining the significance of this property in quantum field theory.

5. Given a bounded operator $\hat{C}$ in a quantum system, explain the significance of its point spectrum compared to its continuous spectrum.

6. Explore the implications of spectral properties on the stability of quantum states within the context of particle interactions.

Answers 1

1. Consider the self-adjoint operator $\hat{A}$ with spectrum $\sigma(\hat{A}) = \{1, 2, 3\}$. Write the spectral decomposition of $\hat{A}$ using projection operators. **Solution:** According to the spectral theorem, a self-adjoint operator can be expressed in terms of its spectrum using projection operators $E(\lambda)$:

$$\hat{A} = \int_{\sigma(\hat{A})} \lambda \, dE(\lambda).$$

   For the discrete spectrum $\sigma(\hat{A}) = \{1, 2, 3\}$, the spectral decomposition is written as:

$$\hat{A} = 1E(1) + 2E(2) + 3E(3),$$

   where $E(i)$ are the projection operators onto the eigenspaces corresponding to each eigenvalue i.

2. Given a Hamiltonian $\hat{H}$ with discrete energy spectrum $\sigma_d(\hat{H}) = \{E_1, E_2, E_3\}$, describe how these energy eigenvalues are related to the physical states of the quantum field theory. **Solution:** In quantum field theory, the Hamiltonian $\hat{H}$ dictates the energy levels of the system. The discrete eigenvalues E_1, E_2, E_3 correspond to stable, quantized energy levels associated with the particle states of the field. Each eigenvalue correlates with an eigenstate $|E_i\rangle$ in the Hilbert space, representing a distinct physical state of the system. The eigenstates form a basis for describing the quantum state of the field, and transitions between these states can be related to particle production or annihilation processes.

3. Demonstrate the application of the spectral representation to a scalar field propagator $D(x - y)$ by expressing it in terms of the spectral density function. **Solution:** The scalar field propagator $D(x - y)$ can be expressed using the spectral representation:

$$D(x - y) = \int_{-\infty}^{+\infty} \frac{d\mu(\sigma)}{2\pi} \, e^{-i\sigma(x-y)},$$

   where $\mu(\sigma)$ is the spectral density function. This representation approaches the evaluation of the propagator by integrating over the possible energy-momentum values distributed according to $\mu(\sigma)$, providing a mathematical framework for predicting field correlations across spacetime events.

4. Prove that a self-adjoint operator $\hat{B}$ has a real spectrum, briefly explaining the significance of this property in quantum field theory. **Solution:** To show that $\hat{B}$ has a real spectrum, consider an eigenvalue equation $\hat{B}|v\rangle = \lambda|v\rangle$ where λ is an eigenvalue and $|v\rangle$ is a corresponding eigenvector. Taking the inner product with itself:

$$\langle v|\hat{B}|v\rangle = \lambda\langle v|v\rangle.$$

   Since $\hat{B}$ is self-adjoint:

$$\langle \hat{B}v|v\rangle = \langle v|\hat{B}|v\rangle = \overline{\langle v|\hat{B}|v\rangle} = \overline{\lambda}\langle v|v\rangle.$$

   Matching both sides, it follows $\lambda = \overline{\lambda} \Rightarrow \lambda$ is real. Real spectra ensure energy observed in quantum systems is a physically meaningful quantity, enabling coherent descriptions of quantum states and their interactions.

5. Given a bounded operator $\hat{C}$ in a quantum system, explain the significance of its point spectrum compared to its continuous spectrum. **Solution:** The point spectrum of a bounded operator $\hat{C}$ consists of isolated eigenvalues, signifying discrete and quantized states within the quantum system, often linked to stable configurations such as bound states or particle states. The continuous spectrum, however, represents a range of energy levels associated with processes like scattering where states are not localized, resembling free particles or extended wave modes. Understanding both spectra informs about the nature and dynamics of quantum processes under different physical conditions.

6. Explore the implications of spectral properties on the stability of quantum states within the context of particle interactions. **Solution:** Spectral properties greatly influence quantum state stability. Stable

169

states are usually associated with eigenvectors linked to bound eigenvalues in the spectrum, reflecting non-decaying, stationary configurations. Conversely, states associated with the continuous spectrum can represent less stable, dispersible or interacting states, potentially leading to decay or transitions due to particle collisions. Evaluating spectral properties aids in predicting particle lifetimes, interaction outcomes, and resonance effects pivotal in characterizing and understanding complex quantum field interactions.

Practice Problems 2

1. Given a self-adjoint operator $\hat{A}$ in a Hilbert space, describe the spectral theorem and derive the integral formulation involving the projection-valued measure $E(\lambda)$.

2. Show that for a Hamiltonian operator $\hat{H}$, which governs the time evolution of a quantum field, the continuous part of the spectrum $\sigma_c(\hat{H})$ is related to scattering states in particle physics.

3. Consider a scalar field operator $\hat{\phi}(x)$ whose Hamiltonian is given by

$$\hat{H} = \int d^3x \, \mathcal{H}(\hat{\phi}(x), \nabla\hat{\phi}(x), \hat{\pi}(x)).$$

   Discuss how the spectral analysis of $\hat{H}$ could determine the mass of particles described by the field $\hat{\phi}(x)$.

4. Explain the importance of the spectral density function $\mu(\sigma)$ in the spectral representation of a field operator propagator $D(x - y)$, and show how it influences the propagator's behavior.

5. Discuss why the eigenvalues of a bounded operator on a Hilbert space are part of its spectrum, and how this differs for unbounded operators such as those found in quantum field theory.

6. Illustrate with an example how the spectral theorem aids in solving differential equations in quantum mechanics, specifically focusing on how it applies to a quantum harmonic oscillator.

Answers 2

1. **Solution:** The spectral theorem for a self-adjoint operator $\hat{A}$ states that $\hat{A}$ can be expressed as:

$$\hat{A} = \int_{\sigma(\hat{A})} \lambda \, dE(\lambda),$$

   where $E(\lambda)$ is a projection-valued measure on the spectrum $\sigma(\hat{A})$. This representation allows us to understand the action of $\hat{A}$ via decomposition into simpler parts related to its eigenvalues and eigenvectors.

2. **Solution:** For a Hamiltonian operator $\hat{H}$, the continuous spectrum $\sigma_c(\hat{H})$ consists of values for which there are no proper eigenfunctions. In particle physics, these values correspond to scattering states where particles are not bound but freely interacting. The continuous spectrum thus characterizes the energies where free particles and their interactions are observed.

3. **Solution:** The spectral properties of a Hamiltonian $\hat{H}$ guide the interpretation of particle masses. When $\hat{H}$ describes a free field, its spectrum includes values representing possible energy levels. These relate to particle masses in quantum field theory (QFT). For example, analyzing the discrete part of the spectrum allows particle mass extraction, where each bound state corresponds to a mass eigenvalue.

4. **Solution:** The spectral density $\mu(\sigma)$ provides critical information on how the spectrum of the operator is populated. In the propagator $D(x - y)$, $\mu(\sigma)$ determines contributions from different energy states to the propagator, affecting how the quantum field evolves over spacetime. This directly influences physical predictions regarding state transitions.

5. **Solution:** In a bounded operator, eigenvalues are countable and form a subset of the spectrum. For unbounded operators like those in QFT, the spectrum can include continuous parts without proper eigenvalues. In such cases, the spectral theorem extends to consider the entire range of possible measurable outcome related to the physical system, including boundary effects at infinity.

6. **Solution:** The spectral theorem's role in solving differential equations exemplifies in the quantum harmonic oscillator problem. The operator corresponding to the Hamiltonian is decomposed into its spectral components, simplifying the equation's solution. Specifically, eigenfunctions (Hermite polynomials here) of the position operator and their associated eigenvalues (energy levels) are used to obtain exact solutions to the harmonic oscillator's Schrödinger equation.

Practice Problems 3

1. Consider a self-adjoint operator $\hat{A}$ acting on a Hilbert space. If its spectrum $\sigma(\hat{A})$ consists only of the discrete spectrum, explain how $\hat{A}$ can be expressed using the spectral theorem. What are the implications of this?

2. Given that the field operator $\hat{\phi}(x)$ evolves according to the Hamiltonian $\hat{H}$, describe the relationship between the spectrum of $\hat{H}$ and the stability of quantum states in a field theory.

3. Describe how the spectral theorem applies to a Hamiltonian operator $\hat{H}$ with a continuous spectrum. What role does the spectral density function play in this context?

4. Explain the significance of the projection-valued measure $E(\lambda)$ in the context of the spectral theorem for an operator $\hat{A}$. How does it relate to the operator's action on quantum states?

5. In the context of particle physics, discuss why it is essential to distinguish between discrete and continuous parts of the spectrum of a Hamiltonian operator. Use examples of particle states to illustrate your point.

6. The spectral representation of a propagator $D(x - y)$ includes the spectral density function $\mu(\sigma)$. Describe this function's role and how it influences particle interaction predictions in quantum field theory.

Answers 3

1. A self-adjoint operator $\hat{A}$ with a spectrum consisting only of discrete eigenvalues is expressed using the spectral theorem as:

$$\hat{A} = \sum_n \lambda_n E_n,$$

where λ_n are the eigenvalues, and E_n are the associated projection operators onto the eigenspaces. The implication is that $\hat{A}$ can be entirely characterized by its eigenvectors and eigenvalues, providing a complete description of the observable it represents.

2. The spectrum of $\hat{H}$ is directly related to the stability of quantum states. Eigenstates corresponding to discrete eigenvalues represent stable or bound states, while continuous spectra often indicate unbound or scattering states. A stable quantum state remains unchanged over time (except for phase), while an unstable state may decay.

3. For a Hamiltonian operator $\hat{H}$ with a continuous spectrum, the spectral theorem uses an integral representation:

$$\hat{H} = \int_{\sigma_c(\hat{H})} \lambda \, dE(\lambda).$$

The spectral density function $\mu(\sigma)$ is crucial for defining the spectrum over the continuous parameter σ. It characterizes the weight of different eigenvalue contributions and is essential for understanding energy distributions in unbound states.

4. The projection-valued measure $E(\lambda)$ essentially decomposes the Hilbert space into invariant subspaces corresponding to different spectral values. For a state $|\psi\rangle$, the action $E(\lambda)|\psi\rangle$ projects this state onto the eigenspace associated with the spectral value λ. It allows for the calculation of expectation values and probabilities.

5. In particle physics, distinguishing between discrete and continuous spectra is vital because bound states, like hadrons, arise from discrete parts of the spectrum representing stable, identifiable particles. The continuous spectrum is related to scattering states, which provide information on how particles interact and transform into one another. For example, the discrete spectrum of the QCD Hamiltonian includes mesons and baryons, while the continuous spectrum relates to multi-particle scattering states.

6. The spectral density function $\mu(\sigma)$ in the spectral representation of $D(x - y)$ influences calculations of probabilities for transitions between states. It dictates the strength and type of interactions as particles propagate through spacetime. Thus, it critically affects predictions of observable phenomena in quantum field theory, like cross-sections for scattering processes.

Chapter 18

Scattering Theory and the S-Matrix

Introduction to Scattering Theory

Scattering theory is integral to the analysis of particle interactions in quantum mechanics and quantum field theory. The core objective is to understand how particles scatter off one another and the associated transition probabilities. Consider a system described by an initial state $|\psi_i\rangle$ that evolves over time into a final state $|\psi_f\rangle$. The ultimate goal is to derive expressions for computing probabilities $P_{i \to f}$ of this transition occurring.

Asymptotic States

In scattering theory, the notion of asymptotic states plays a central role. These states describe the behavior of particles in the distant past and future, free from interaction effects. The **in** state $|\psi_{\text{in}}\rangle$ and the **out** state $|\psi_{\text{out}}\rangle$ are asymptotic states that are free fields when $t \to \pm\infty$.

Define the asymptotic states as:

$$|\psi_{\text{in}}\rangle = \lim_{t \to -\infty} e^{iHt} e^{-iH_0 t} |\phi\rangle, \quad |\psi_{\text{out}}\rangle = \lim_{t \to \infty} e^{iHt} e^{-iH_0 t} |\phi\rangle,$$

where H is the full Hamiltonian and H_0 is the free Hamiltonian.

The S-Matrix

The S-matrix, or scattering matrix, encapsulates information about the scattering process. It relates the **in** and **out** states through:

$$|\psi_{\text{out}}\rangle = S|\psi_{\text{in}}\rangle.$$

The S-matrix elements S_{fi} are given by the inner product:

$$S_{fi} = \langle \psi_{\text{out}}^f | \psi_{\text{in}}^i \rangle,$$

where $|\psi_{\text{out}}^f\rangle$ and $|\psi_{\text{in}}^i\rangle$ are specific asymptotic states corresponding to final and initial configurations, respectively.

The connection between the time evolution operator $U(t, t_0)$ and the S-matrix is key. For $t_0 \to -\infty$ and $t \to \infty$, the limit exists:

$$S = \lim_{t_0 \to -\infty, t \to \infty} U(t, t_0).$$

Lippmann-Schwinger Equation

The Lippmann-Schwinger equation is foundational for understanding the perturbative framework of scattering theory. It symbolically expresses the relationship of the **in** and **out** states:

$$|\psi_f^\pm\rangle = |\psi_i\rangle + \frac{1}{E - H_0 \pm i\epsilon} V |\psi_f^\pm\rangle,$$

where V is the interaction term, and E is the energy of the system. The $\pm i\epsilon$ ensures the boundary conditions corresponding to states evolving forward or backward in time.

Mathematical Formulation of Scattering Processes

Considering the above frameworks, scattering processes are reformulated mathematically with an emphasis on operator theory. Transition probabilities are derived from the modulus squared of the S-matrix elements:

$$P_{i \to f} = |S_{fi}|^2.$$

Within the Hilbert space structure, the completeness and unitarity of the S-matrix ensure that the total probability is conserved:

$$SS^\dagger = S^\dagger S = I.$$

Applications of the S-Matrix

Within quantum field theory, the S-matrix formalism extends beyond elementary particles to complex systems, including multi-particle interactions and field excitations. Employing perturbation theory, the S-matrix elements are systematically approximated using Feynman diagrams. This provides calculable predictions for experimentally measurable quantities, such as cross sections and decay rates.

Practice Problems 1

1. Prove that the S-matrix is unitary, i.e., show that $SS^\dagger = S^\dagger S = I$.

2. Derive the expression for the transition probability $P_{i \to f}$ in terms of the S-matrix elements.

3. Explain the significance of the asymptotic state condition $|\psi_{\text{in/out}}\rangle = \lim_{t \to \mp \infty} e^{iHt} e^{-iH_0 t} |\phi\rangle$.

4. Provide the derivation of the Lippmann-Schwinger equation $|\psi_f^{\pm}\rangle = |\psi_i\rangle + \frac{1}{E - H_0 \pm i\epsilon} V |\psi_f^{\pm}\rangle$.

5. Considering two asymptotic states $|\psi_{\text{in}}\rangle$ and $|\psi_{\text{out}}\rangle$, show how they are related through the S-matrix S.

6. Discuss the role of Feynman diagrams in calculating the S-matrix elements for multi-particle interactions.

Answers 1

1. **Solution:**

The S-matrix is unitary if $SS^\dagger = S^\dagger S = I$. This implies conservation of probability:

$$\int d\psi_{\text{out}} |S\psi_{\text{in}}|^2 = \int d\psi_{\text{in}} |\psi_{\text{in}}|^2.$$

Since the time evolution operator $U(t, t_0)$ is unitary for finite times, and $S = \lim_{t_0 \to -\infty, t \to \infty} U(t, t_0)$, unitarity extends under this limit.

Therefore,

$$(SS^\dagger)_{jf} = \sum_i S_{ji} S_{fi}^* = \delta_{jf},$$

which implies $SS^\dagger = I$, showing that the S-matrix is unitary.

2. **Solution:**

The transition probability from the initial state $|\psi_{\text{in}}^i\rangle$ to the final state $|\psi_{\text{out}}^f\rangle$ is given by the modulus squared of the S-matrix element:

$$P_{i \to f} = |S_{fi}|^2 = |\langle \psi_{\text{out}}^f | S | \psi_{\text{in}}^i \rangle|^2.$$

Here, S_{fi} is the element of the S-matrix that connects the specific initial and final asymptotic states.

3. **Solution:**

The asymptotic states formalize the idea that particles are free before and after interaction. The condition

$$|\psi_{\text{in/out}}\rangle = \lim_{t \to \mp\infty} e^{iHt} e^{-iH_0 t} |\phi\rangle$$

ensures that as $t \to \pm\infty$, the interactions vanish, thus the states act as free states evolving under H_0. This framework simplifies handling interactions through asymptotic completeness.

4. **Solution:**

The Lippmann-Schwinger equation derives from the perturbative expansion of the quantum state $|\psi\rangle$ in terms of free states $|\phi\rangle$.

Begin with:

$$|\psi\rangle = |\phi\rangle + \frac{1}{E - H \pm i\epsilon} V |\psi\rangle.$$

Substitute $H = H_0 + V$:

$$|\psi_f^\pm\rangle = |\psi_i\rangle + \frac{1}{E - H_0 \pm i\epsilon} V |\psi_f^\pm\rangle.$$

This equation accounts for forward and backward evolving states, with $i\epsilon$ guaranteeing convergence based on causality.

5. **Solution:**

As prior discussed, $|\psi_{\text{out}}\rangle = S|\psi_{\text{in}}\rangle$, which denotes:

$$S_{fi} = \langle \psi_{\text{out}}^f | \psi_{\text{in}}^i \rangle.$$

This clearly draws a direct correspondence between initial and final asymptotic states via the S-matrix, expressing transitions in an interaction-dominated region in terms of free regions.

6. **Solution:**

Feynman diagrams serve as a visual representation of perturbative expansion terms when calculating S-matrix elements. Each vertex and line corresponds to specific interaction terms and propagators.

In essence, the diagrams act as a bookkeeping method for terms contributing to scattering amplitudes, allowing computation of cross sections and decay rates based on field interactions described by the underlying quantum mechanics principle.

Practice Problems 2

1. Demonstrate how to derive the expression for the S-matrix elements S_{fi} from the asymptotic states $|\psi_{in}^i\rangle$ and $|\psi_{out}^f\rangle$ using bra-ket notation.

2. Explain how the Lippmann-Schwinger equation is used to express the **out** state as a perturbation expansion. Provide the first-order correction term.

3. Show the unitarity condition of the S-matrix and how it ensures the conservation of probability in scattering processes. Demonstrate this using the relationship $SS^\dagger = I$.

4. Analyze how the S-matrix is connected to the time evolution operator $U(t, t_0)$ in the interaction picture. Explain the significance of the limit $t_0 \to -\infty$ and $t \to \infty$.

5. Derive the relationship between the transition probability $P_{i \to f}$ and the S-matrix element S_{fi}. Why is this important in the context of quantum field theory?

6. Discuss the role of Feynman diagrams in calculating S-matrix elements. How do they simplify the complex calculations involved in scattering processes?

Answers 2

1. **Solution:** The S-matrix elements S_{fi} relate the asymptotic **in** and **out** states. The expression is:

$$S_{fi} = \langle \psi_{\text{out}}^f | \psi_{\text{in}}^i \rangle.$$

This inner product evaluates the overlap between the final **out** state and the initial **in** state, crucial for calculating the transition amplitude from the initial to the final state in scattering processes.

2. **Solution:** The Lippmann-Schwinger equation for the **out** state is:

$$|\psi_f^+\rangle = |\psi_i\rangle + \frac{1}{E - H_0 + i\epsilon} V |\psi_f^+\rangle.$$

Expanding perturbatively, the first-order correction term is:

$$|\psi_f^{(1)+}\rangle = \frac{1}{E - H_0 + i\epsilon} V |\psi_i\rangle.$$

This equation provides a way to compute corrections to the free state due to interactions.

3. **Solution:** The unitarity condition of the S-matrix is $SS^\dagger = S^\dagger S = I$, ensuring that the total probability is conserved. This means that any probability lost to the initial configuration is accounted for in the final state. Mathematically, this implies that:

$$\sum_f |S_{fi}|^2 = 1,$$

reflecting the conservation of the overall probability amplitude in scattering events.

4. **Solution:** In the interaction picture, the S-matrix is linked to the time evolution operator by:

$$S = \lim_{t_0 \to -\infty, t \to \infty} U(t, t_0).$$

This limit isolates free particle dynamics from interactions over infinite time, simplifying calculation by explicitly focusing on interaction effects.

5. **Solution:** The transition probability $P_{i \to f}$ is the modulus square of the S-matrix element:

$$P_{i \to f} = |S_{fi}|^2.$$

This quantifies how likely the transition from state i to state f is, according to the dynamics embodied in the S-matrix. It's essential for predictive calculations in quantum field theory.

6. **Solution:** Feynman diagrams represent terms in the perturbative expansion of S-matrix elements. These diagrams provide a visual shorthand for complex integrals over particle interactions, organizing calculations systematically. They simplify accounting for all possible interaction pathways in quantum field processes and offer computational techniques for scattering amplitudes efficiently.

Practice Problems 3

1. Show that the S-matrix S is unitary, i.e., prove that $SS^\dagger = I$.

2. Explain the physical significance of the asymptotic states $|\psi_{\text{in}}\rangle$ and $|\psi_{\text{out}}\rangle$ in scattering theory.

3. Derive the expression for the S-matrix S using the Lippmann-Schwinger equation for a potential V.

4. Discuss the significance of the operator $U(t, t_0)$ in the context of the S-matrix and its relation with time evolution.

5. Show how the probability of transition $P_{i \to f}$ can be expressed in terms of the S-matrix elements S_{fi}.

6. Explain how Feynman diagrams are used in conjunction with the S-matrix formalism to compute scattering amplitudes.

Answers 3

1. Show that the S-matrix S is unitary, i.e., prove that $SS^\dagger = I$.

   Solution:
   To prove that the S-matrix is unitary, we need to show that $SS^\dagger = I$. The S-matrix relates the **in** and **out** states via $|\psi_{\text{out}}\rangle = S|\psi_{\text{in}}\rangle$. For unitarity, we have:

   $$\langle \psi_{\text{out}}^f | \psi_{\text{out}}^i \rangle = \langle \psi_{\text{in}}^f | \psi_{\text{in}}^i \rangle.$$

   Since $|\psi_{\text{out}}^f\rangle = S|\psi_{\text{in}}^f\rangle$ and $|\psi_{\text{out}}^i\rangle = S|\psi_{\text{in}}^i\rangle$, it follows:

   $$\langle \psi_{\text{in}}^f | S^\dagger S | \psi_{\text{in}}^i \rangle = \langle \psi_{\text{in}}^f | \psi_{\text{in}}^i \rangle.$$

   This implies $S^\dagger S = I$. Thus, unitarity holds for the S-matrix.

2. Explain the physical significance of the asymptotic states $|\psi_{\text{in}}\rangle$ and $|\psi_{\text{out}}\rangle$ in scattering theory.

   Solution:
   Asymptotic states, $|\psi_{\text{in}}\rangle$ and $|\psi_{\text{out}}\rangle$, represent the states of a quantum system in the distant past and future where interactions are negligible. In other words, these states correspond to free particles moving before and after an interaction occurs. The $|\psi_{\text{in}}\rangle$ state describes incoming particles approaching the interaction region, while $|\psi_{\text{out}}\rangle$ reflects outgoing particles that have already undergone interaction. They are key to defining the S-matrix and computing scattering probabilities.

3. Derive the expression for the S-matrix S using the Lippmann-Schwinger equation for a potential V.

   Solution:
   The Lippmann-Schwinger equation relates scattering states to the potential:

   $$|\psi^\pm\rangle = |\phi\rangle + \frac{1}{E - H_0 \pm i\epsilon} V |\psi^\pm\rangle.$$

   Multiplying by $\langle\phi|$ and solving for $S_{fi} = \langle\phi_f|\psi^+\rangle$, we obtain:

   $$S_{fi} = \langle\phi_f| \left(1 + \frac{1}{E - H_0 + i\epsilon} V \right) |\phi_i\rangle.$$

   Thus, the S-matrix near the energy shell is captured by integral terms involving potential V, enabling perturbative calculations.

4. Discuss the significance of the operator $U(t, t_0)$ in the context of the S-matrix and its relation with time evolution.

   Solution:
   The operator $U(t, t_0)$, the time evolution operator, dictates how quantum states develop between initial time t_0 and final time t. It connects the **in** and **out** states:

   $$|\psi(t)\rangle = U(t, t_0)|\psi(t_0)\rangle.$$

The S-matrix is the limiting case as $t_0 \to -\infty$ and $t \to \infty$:

$$S = \lim_{t_0 \to -\infty, t \to \infty} U(t, t_0).$$

It thereby encapsulates the entire interaction history in scattering processes through asymptotic limits.

5. Show how the probability of transition $P_{i \to f}$ can be expressed in terms of the S-matrix elements S_{fi}.

Solution:
The transition probability from an initial state $|\psi_{\text{in}}^i\rangle$ to a final state $|\psi_{\text{out}}^f\rangle$ is given by the square of the amplitude of the corresponding S-matrix element:

$$P_{i \to f} = |S_{fi}|^2.$$

This expression arises from the probability interpretation of quantum mechanics where measurements are tied to the modulus squared of amplitude functions. Hence, it allows us to compute seesible outcomes from theoretical scattering models.

6. Explain how Feynman diagrams are used in conjunction with the S-matrix formalism to compute scattering amplitudes.

Solution:
Feynman diagrams provide a visual and computational tool for evaluating S-matrix elements perturbatively. Each diagram represents a particular term in the series expansion of the exponential evolution operator in interaction picture:

$$e^{-iHt} = T\left\{ \exp\left(-i \int_{t_0}^{t} H_I(t') dt' \right) \right\}.$$

Lines and vertices correspond to propagators and interaction vertices in the expanded series, corresponding to integrals over spacetime that collectively construct scattering amplitudes. Thus, Feynman diagrams serve as intermediary steps toward computing S_{fi} and related quantities like cross sections and decay rates.

Chapter 19

Wightman Axioms and Mathematical Foundations of QFT

Introduction to Wightman Axioms

The Wightman axioms provide a rigorous framework for quantum field theory (QFT) by establishing a set of conditions that quantum fields must satisfy. These axioms were designed to ensure both the mathematical consistency and the physical relevance of QFT.

The axioms formalize the properties that quantum fields must exhibit and are typically defined in the context of a Hilbert space $\mathcal{H}$ with a time-evolution represented by a unitary operator. Fields are operator-valued distributions, denoted typically by $\phi(x)$.

Axiom I: Existence of Hilbert Space

The foundational space in QFT is a complex Hilbert space $\mathcal{H}$. This space is equipped with an inner product that satisfies:

$$\langle \psi | \phi \rangle = \overline{\langle \phi | \psi \rangle}$$

for states $|\psi\rangle, |\phi\rangle \in \mathcal{H}$. The state space contains a vacuum state $|\Omega\rangle$, and all physical states are represented as vectors in this space.

Axiom II: Field Operators

Fields are introduced as operator-valued distributions $\phi(x)$ on space-time. The action of a field operator is to create or annihilate particles at each point in space-time. These operators are typically tempered distributions, meaning they map test functions from a Schwartz space $\mathcal{S}$ to the Hilbert space $\mathcal{H}$.

The locality axiom specifies that these fields commute (or anti-commute for fermionic fields) at spacelike separations:

$$[\phi(x), \phi(y)] = 0, \quad \text{for } (x - y)^2 < 0.$$

Axiom III: Covariance

Covariance implies that the field equations are invariant under the Poincaré group, the symmetry group of Minkowski spacetime. For each element (a, Λ) in the Poincaré group, there exists a unitary operator $U(a, \Lambda)$ such that:

$$U(a, \Lambda)\phi(x)U(a, \Lambda)^{-1} = \phi(\Lambda x + a).$$

This ensures that physical predictions do not depend on the observer's inertial frame.

Axiom IV: Spectrum Condition

This axiom ensures that energy and momentum are non-negative. The spectrum of the energy-momentum operator (P^μ) is contained within the future light cone, specifically:

$$\text{Spec}(P^\mu) \subset \{p \in \mathbb{R}^4 \mid p^0 \geq 0\}.$$

It guarantees that the energy observed in any quantum state does not decrease below the vacuum state, reflecting physical stability.

Axiom V: Uniqueness of the Vacuum

The vacuum $|\Omega\rangle$ is invariant under translations by the energy-momentum operator:

$$U(a, I)|\Omega\rangle = |\Omega\rangle.$$

This condition ensures the homogeneity of space-time in QFT, as the vacuum state appears the same from any reference frame.

Construction of Quantum Fields

Constructing quantum fields rigorously involves defining a net of local algebras of observable operators $\mathcal{A}(\mathcal{O})$ on the Hilbert space $\mathcal{H}$, associated with each open region $\mathcal{O}$ in space-time. These algebras accommodate operations that generate local physical quantities.

The Haag-Kastler axioms also play a role in structuring quantum fields by using C*-algebras to describe the algebra of observables. The algebras must satisfy isotony, locality, covariance, and additivity conditions.

Ensuring Mathematical Rigor in QFT

To uphold mathematical rigor, the Wightman axioms impose restrictions on fields as operator-valued distributions rather than conventional functions. This accounts for the indefiniteness in value at each point – a fundamental aspect when considering interactions at quantum scale.

The use of distributions allows the extension of classical solutions in a mathematically precise manner, facilitating the application of techniques from functional analysis and operator theory to tackle perturbation series and renormalization. This yields a mathematically concise formulation that aligns with experimental observations in particle physics without ambiguity in definitions or computations.

Practice Problems 1

1. Discuss the significance of the vacuum state $|\Omega\rangle$ in the Wightman Axioms and describe its mathematical properties.

2. Explain the role of operator-valued distributions in formulating quantum fields, using the context provided by the Wightman Axioms.

3. What is the spectrum condition in the Wightman axioms and how does it relate to the physical stability of quantum systems?

4. How do the Wightman axioms ensure the covariance of quantum fields under transformations? Provide a detailed explanation.

5. Describe how the locality axiom in the Wightman framework impacts the commutation relations between quantum fields.

6. Illustrate the use of distributions in quantum field theory for addressing interactions at the quantum scale as per the Wightman axioms.

Answers 1

1. Discuss the significance of the vacuum state $|\Omega\rangle$ in the Wightman Axioms and describe its mathematical properties.
   Solution:

   The vacuum state $|\Omega\rangle$ in the Wightman Axioms serves as the ground state of the quantum field theory, where no particles are present. It is crucial for several reasons:

   - **Uniqueness and Invariance:** The vacuum is unique and invariant under translations, providing a stable reference point in the Hilbert space $\mathcal{H}$.
   - **Energy Minimization:** As the state with the lowest energy, it ensures that all physical states have energy levels non-lower than the vacuum, adhering to the spectrum condition.
   - **Poincaré Invariance:** Ensures symmetry under the Poincaré group, playing a role in the covariance axiom.

   Thus, $|\Omega\rangle$ facilitates the mathematical structure aligned with physical principles.

2. Explain the role of operator-valued distributions in formulating quantum fields, using the context provided by the Wightman Axioms.
   Solution:

   Operator-valued distributions, such as $\phi(x)$, extend the concept of fields to a mathematically consistent framework in quantum theory.

   - **Distributions:** They map test functions from Schwartz space $\mathcal{S}$ to the Hilbert space $\mathcal{H}$, allowing flexible treatment of pointwise interactions.
   - **Locality:** Facilitate the locality axiom by operating within open regions of spacetime, where their commutation relations are enforced at spacelike separations.
   - **Regularization:** Help handle infinities and undefined products that arise in quantum field interactions through regularization techniques.

   This formulation is critical for tackling the infinite degrees of freedom in QFT.

3. What is the spectrum condition in the Wightman axioms and how does it relate to the physical stability of quantum systems?
   Solution:

   The spectrum condition demands that the spectrum of the energy-momentum operator (P^μ) lies within the future light cone $(p^0 \geq 0)$.

   - **Physical Stability:** Ensures non-negative energy, reflecting that energy cannot decrease below the vacuum state; crucial for stable quantum systems.
   - **Causality Compliance:** Acts in tandem with causality principles by restricting the propagation of signals outside the light cone.

   It maintains realistic energy dynamics in quantum theories.

4. How do the Wightman axioms ensure the covariance of quantum fields under transformations? Provide a detailed explanation.
   Solution:

   Covariance in the Wightman axioms is achieved by field operator transformation under the Poincaré group:
   $$U(a, \Lambda)\phi(x)U(a, \Lambda)^{-1} = \phi(\Lambda x + a)$$

   - **Invariant Physical Laws:** Poincaré transformations (translations and Lorentz transformations) reflect uniform physical laws across spacetime.

- **Unitary Evolution:** The unitary operator $U(a, \Lambda)$ ensures that field states evolve without altering inner product structure, preserving probabilities.

This principle aligns with relativity, ensuring all observers detect identical physics.

5. Describe how the locality axiom in the Wightman framework impacts the commutation relations between quantum fields.
   Solution:

   The locality axiom states that fields at spacelike separations must commute (or anti-commute for fermions):
   $$[\phi(x), \phi(y)] = 0, \ (x - y)^2 < 0$$

   - **Causality:** Reflects causal independence between spatially separated events; no signal travels faster than light.
   - **Probability Conservation:** Local observables uphold probabilities unaffected by remote measurements, conserving information flow integrity.

   This supports robust quantum mechanical models integrating relativity.

6. Illustrate the use of distributions in quantum field theory for addressing interactions at the quantum scale as per the Wightman axioms.
   Solution:

   The use of distributions permits handling of quantum scale interactions with precision.

   - **Singular Behavior Resolution:** Treats the singularities at points of contact where functions are traditionally undefined.
   - **Renormalization Compatibility:** Facilitates regularization and renormalization, crucial for theoretical predictions aligning with experimental results.
   - **Functional Analysis Application:** Drifts problems into a well-defined mathematical regime, allowing robust techniques of functional analysis to be implemented.

   This method overcomes challenges inherent in quantizing fields accurately.

Practice Problems 2

1. Explain the concept of operator-valued distributions in the framework of the Wightman axioms. Why are fields considered as distributions rather than functions?

2. Discuss the significance of the commutation or anticommutation relations for field operators at spacelike intervals. How do these relations contribute to the locality condition in quantum field theory?

3. Illustrate the role of the Poincaré group in the Wightman axioms. How does covariance under the Poincaré group ensure the invariance of field equations?

4. Define the spectrum condition within the context of the Wightman axioms. How does this condition relate to the physical interpretation of energy in quantum field theory?

5. Describe the construction of quantum fields through local algebras of observables. How do the Haag-Kastler axioms relate to the organization of these algebras?

6. How does mathematical rigor, particularly through the use of distributions, facilitate overcoming challenges in quantum field theory like perturbation series and renormalization?

Answers 2

1. **Operator-valued Distributions:**

   Operator-valued distributions are used in quantum field theory to handle fields that cannot be treated adequately as functions because they can exhibit singular behavior at certain points. A distribution, in this context, is a generalized function that acts on a set of test functions from a Schwartz space $\mathcal{S}$. This approach allows the mathematical framework to account for singularities and non-localizable properties of quantum fields. By using distributions, we avoid trying to assign definite values at each point in spacetime, which is critical given the intrinsic uncertainties and divergences in quantum fields. This treatment aligns with the requirement of fields being tempered distributions, allowing the application of advanced techniques from functional analysis.

2. **Commutation Relations and Locality:**

The commutation or anticommutation relations of field operators ensure the principle of locality in quantum field theory. Specifically, for spacelike separations $(x - y)^2 < 0$, field operators must satisfy the relations $[\phi(x), \phi(y)] = 0$ for bosons and $\{\phi(x), \phi(y)\} = 0$ for fermions. This implies that operations performed at spacelike-separated points cannot affect each other, maintaining causality in accord with the principles of relativity. Such relations are central to ensuring that quantum fields respect the speed-of-light constraint on information propagation.

3. **Poincaré Group and Covariance:**

The Poincaré group embodies the symmetries of Minkowski spacetime, including translations, rotations, and boosts. Covariance under the Poincaré group indicates that the laws of physics, represented by field equations, remain invariant for different inertial observers. This is mathematically expressed as $U(a, \Lambda)\phi(x)U(a, \Lambda)^{-1} = \phi(\Lambda x + a)$, where $U(a, \Lambda)$ are unitary operators associated with the group elements. It ensures that physical predictions are independent of the observer's frame of reference, preserving the physically observed symmetries of spacetime.

4. **Spectrum Condition:**

The spectrum condition restricts the spectrum of the energy-momentum operator (P^μ) to lie within the future light cone, specifically requiring that $\text{Spec}(P^\mu) \subset \{p \in \mathbb{R}^4 \mid p^0 \geq 0\}$. This condition reflects the physical stability of quantum systems by ensuring that energy cannot be negative, aligning with the concept that the vacuum state has the lowest possible energy. It provides a consistent framework where the direction of time is preserved, and the evolution of states maintains causality.

5. **Quantum Field Construction via Local Algebras:**

Quantum fields are constructed using a net of local algebras of observables $\mathcal{A}(\mathcal{O})$, defined for each open region in spacetime. The Haag-Kastler axioms specify properties like isotony (ordering of regions implies ordering of algebras), locality, covariance, and additivity, which organize these algebras. These axioms facilitate a clear representation of local physical quantities and their interrelationships, allowing the rigorous treatment of quantum states and observables in a mathematically structured layout.

6. **Mathematical Rigor through Distributions:**

Employing distributions in quantum field theory allows for a mathematically rigorous treatment of scenarios where traditional functions fail due to singularities and infinities. This rigor is crucial for addressing challenges in perturbation series and renormalization where formal series and divergences need precise mathematical handling. Through the use of functional analysis and operator theory, distributions enable the extension of classical results into the quantum regime, supporting calculations that align with experimental realities while maintaining theoretical soundness.

Practice Problems 3

1. Discuss the significance of the Hilbert space in the Wightman axioms and its fundamental properties.

2. Explain how operator-valued distributions differ from conventional functions and their importance in QFT.

3. Define spacelike separation and elaborate on its role in the locality axiom as applied to quantum fields.

4. Illustrate the concept of covariance in QFT and demonstrate this with an example involving a transformation of field operators.

5. Detail the spectrum condition and analyze its implications for the stability of quantum systems.

6. Delve into the construction of quantum fields using C*-algebras and their relevance in ensuring mathematical rigor.

Answers 3

1. Discuss the significance of the Hilbert space in the Wightman axioms and its fundamental properties.
   Solution:

   - The Hilbert space $\mathcal{H}$ is crucial as it serves as the foundation for describing quantum states in QFT. It provides a complete mathematical framework that includes an inner product, enabling the calculation of probabilities and other physical quantities.
   - The inner product $\langle \psi | \phi \rangle = \overline{\langle \phi | \psi \rangle}$ establishes the quantum mechanical probability interpretation through normalization and orthogonality conditions.
   - Hilbert spaces ensure that all physical states are represented as vectors, and it contains the vacuum state $|\Omega\rangle$, which serves as the base state for constructing other states.
   - Completeness of the space ensures convergence, critical for defining limits and performing functional operations, crucial in the analysis of quantum phenomena.

2. Explain how operator-valued distributions differ from conventional functions and their importance in QFT.
   Solution:

   - Operator-valued distributions like $\phi(x)$ are not functions in the classical sense; they assign operators rather than scalars to points.
   - These distributions operate over test functions within a Schwartz space $\mathcal{S}$, allowing singularities and infinities typically encountered at quantum scales to be managed rigorously.
   - They encapsulate the probabilistic nature of quantum field interactions, crucial for the application of perturbative and non-perturbative analysis.
   - Their distributional nature allows the reconciliation of theoretical constructs with physical predictions and experimental data.

3. Define spacelike separation and elaborate on its role in the locality axiom as applied to quantum fields.
   Solution:

   - Spacelike separation between two points x, y means $(x - y)^2 < 0$, indicating that no signal can travel faster than light between these points.
   - In the locality axiom, it ensures non-interaction or independence, such that $[\phi(x), \phi(y)] = 0$ or $\{\phi(x), \phi(y)\} = 0$ for fermions, meaning measurements at spacelike intervals do not affect each other.
   - This axiom upholds causality, a core principle in relativity, ensuring that the order of observations does not impart information paradoxically.
   - The mathematical formulation directly conveys stability, allowing the logical structuring of field theories applicable to particle collision and interactions.

4. Illustrate the concept of covariance in QFT and demonstrate this with an example involving a transformation of field operators.
   Solution:

   - Covariance in QFT means formulating theories that remain invariant under the Poincaré group transformations.
   - Fields transform with $U(a, \Lambda)\phi(x)U(a, \Lambda)^{-1} = \phi(\Lambda x + a)$, representing translation and Lorentz boosts.
   - **Example:** Consider $\phi(x) \rightarrow U(a, \Lambda)\phi(x) = \phi'(\Lambda x + a)$; it retains the field algebra but in a transformed, physically equivalent state.
   - By ensuring covariant transformation, the physics modeled is consistent across different frames, essential for ensuring relativistic invariant predictions.

191

5. Detail the spectrum condition and analyze its implications for the stability of quantum systems.
   Solution:

   - The spectrum condition posits the energy-momentum spectrum (P^μ) lies within or on the future light cone $p^0 \geq 0$, ensuring positive energy states.

   - Preventing negative energy prevents unphysical states, thus maintaining system stability essential for defining grounded quantum mechanics.

   - From a practical viewpoint, the condition aligns with observed physical phenomena where energy conservation and causality are upheld.

   - Analyzing systems under this condition affirms consistent formulation of quantum states that adhere to known fundamental laws of physics.

6. Delve into the construction of quantum fields using C*-algebras and their relevance in ensuring mathematical rigor.
   Solution:

   - C*-algebras provide an abstract algebraic framework encompassing observable operators $\mathcal{A}(\mathcal{O})$ within open regions $\mathcal{O}$ in spacetime.

   - Utilizing C*-algebras enables the precise definition of quantum fields through isotony, additivity, covariance, and locality, crucial for algebraic QFT.

   - Mathematical rigor is further ensured by accommodating infinite-dimensional representation, allowing seamless interactions and calculations consistent with physical observables.

   - These construction techniques avoid ambiguities in field operations, hardening the theoretical basis against known quantum phenomena and aligning with experimentative results.

Chapter 20

Haag-Ruelle Theory

Foundations of Haag-Ruelle Theory

Haag-Ruelle theory provides a formulation for scattering processes within the framework of quantum field theory (QFT). It builds on the axiomatic approach of quantum fields, particularly leveraging the properties defined by the Wightman axioms. The primary goal is to establish conditions under which asymptotic fields exist, facilitating an understanding of how particles scatter.

Consider a quantum field $\phi(x)$ acting on a Hilbert space $\mathcal{H}$. The scattering states are constructed in terms of free fields, denoted by $\phi_{in}(x)$ and $\phi_{out}(x)$, representing incoming and outgoing states, respectively. These asymptotic fields are formally defined as:

$$\phi_{in}(x) = \lim_{t \to -\infty} e^{iHt} \phi(x) e^{-iHt}, \tag{20.1}$$

$$\phi_{out}(x) = \lim_{t \to +\infty} e^{iHt} \phi(x) e^{-iHt}, \tag{20.2}$$

where H is the Hamiltonian operator dictating the system's time evolution.

Asymptotic Completeness

Asymptotic completeness is a central concept in scattering theory, expressing that the Hilbert space $\mathcal{H}$ of states can be decomposed into the direct sum of incoming and outgoing Fock spaces:

$$\mathcal{H} = \mathcal{H}_{in} \oplus \mathcal{H}_{out}. \tag{20.3}$$

It asserts that every state in $\mathcal{H}$ can be uniquely described in terms of asymptotic free states, facilitating meaningful scattering interpretations.

Scattering amplitudes are computed using the overlap between ϕ_{in} and ϕ_{out}, which are connected by the S-Matrix S. The matrix elements of S are expressed as:

$$\langle \phi_{out} | S | \phi_{in} \rangle. \tag{20.4}$$

These elements encode the probability amplitudes for transitions between asymptotic states.

Haag's Theorem

Haag's theorem presents a fundamental challenge to the canonical field quantization in interaction representations. It asserts that under reasonable assumptions, such as the presence of a single Hilbert space for both interacting and free fields, the field operators cannot be both correct representations of physical asymptotic fields and satisfy commutation relations.

The implication is that one must carefully construct interacting fields distinct from asymptotic free fields, reflecting genuine dynamics in the quantum field.

Ruelle's Formulation

Building on Haag's insights, Ruelle developed a formal framework for scattering in QFT circumventing Haag's theorem's limitations. The approach relies on localized observables and operators to define out-states distinctly from in-states within localized spacetime regions.

Ruelle's formulation asserts that the existence of asymptotic fields depends on various fundamental limits and correlations evaluated over these regions:

$$\lim_{t \to \pm\infty} \langle \psi | \phi_{out}(t) \phi_{in}(-t) | \psi \rangle < \infty. \tag{20.5}$$

This condition describes states scattering with finite probability, linked to constraints imposed by locality and micro-causality.

Roles in Quantum Field Theory

Haag-Ruelle theory elucidates the mechanism of particle interactions fundamentally driven by underlying field dynamics. It forms the bedrock for understanding non-perturbative phenomena by allowing asymptotic analysis of interacting quantum fields.

Through this theory, consistent frameworks can be constructed to model physical systems where particle creation and annihilation are critically governed by field properties, adhering to fundamental principles such as causality, locality, and covariance.

Researchers often extend Haag-Ruelle results to handle complex interactions involving fields with higher spin or systems under non-trivial external conditions, ultimately refining the grasp on how field theories describe observable phenomena within an experimentally verifiable domain.

Practice Problems 1

1. Consider a quantum field $\phi(x)$ in a Hilbert space $\mathcal{H}$. Show that the asymptotic field $\phi_{in}(x)$ as $t \to -\infty$ can be expressed using the exponential of the Hamiltonian operator H. Provide the formal definition and verify its formulation.

2. Explain the concept of asymptotic completeness in the context of Haag-Ruelle theory. Demonstrate how the Hilbert space $\mathcal{H}$ is decomposed into $\mathcal{H}_{in}$ and $\mathcal{H}_{out}$.

3. Derive the expression for the matrix elements of the S-matrix S given the overlap between the asymptotic fields $\phi_{in}(x)$ and $\phi_{out}(x)$.

4. Discuss Haag's theorem in the framework of QFT, and explain how it impacts the representation of field operators in the interaction picture.

5. Elucidate Ruelle's formal framework for handling scattering in QFT, particularly focusing on the limitations presented by Haag's theorem.

6. Analyze the significance of Haag-Ruelle theory in understanding non-perturbative phenomena in quantum field theory. How does it underlie the consistent modeling of particle interactions?

Answers 1

1. Consider a quantum field $\phi(x)$ in a Hilbert space $\mathcal{H}$. Show that the asymptotic field $\phi_{in}(x)$ as $t \to -\infty$ can be expressed using the exponential of the Hamiltonian operator H. Provide the formal definition

and verify its formulation.

Solution: The asymptotic field $\phi_{in}(x)$ is defined as:

$$\phi_{in}(x) = \lim_{t \to -\infty} e^{iHt}\phi(x)e^{-iHt},$$

where H is the Hamiltonian operator. This definition arises from the notion that, at very large negative times, the evolution of the field should approximate that of a free field. The exponential operator $e^{\pm iHt}$ describes the time evolution dictated by H, and thus asymptotically brings the interacting field $\phi(x)$ to resemble an incoming free field configuration. Verification involves acknowledging its derivation from the Wightman framework and principles such as micro-causality and covariance.

2. Explain the concept of asymptotic completeness in the context of Haag-Ruelle theory. Demonstrate how the Hilbert space $\mathcal{H}$ is decomposed into $\mathcal{H}_{in}$ and $\mathcal{H}_{out}$.

    Solution: Asymptotic completeness asserts that the total Hilbert space $\mathcal{H}$ is the direct sum of the incoming $\mathcal{H}_{in}$ and outgoing $\mathcal{H}_{out}$ Fock spaces:

$$\mathcal{H} = \mathcal{H}_{in} \oplus \mathcal{H}_{out}.$$

    This means that every physical state can be described as a combination of incoming and outgoing asymptotic states. It is essential for interpreting scattering processes since it ensures that all interaction processes start and end in states that can be treated perturbatively. The proof involves showing that interactions can be thus "disentangled" at infinite times, consistently representing physical observations.

3. Derive the expression for the matrix elements of the S-matrix S given the overlap between the asymptotic fields $\phi_{in}(x)$ and $\phi_{out}(x)$.

    Solution: The S-matrix elements are given by:

$$\langle \phi_{out} | S | \phi_{in} \rangle.$$

    For these to be meaningful, an overlap between the asymptotic fields' states must be established. These elements represent the probability amplitude for the transition from an incoming state to an outgoing state. The derivation involves constructing in-states and out-states via the asymptotic fields and their respective time evolution, using the S-matrix as a tool to connect these states in a quantum field theoretic context.

4. Discuss Haag's theorem in the framework of QFT, and explain how it impacts the representation of field operators in the interaction picture.

    Solution: Haag's theorem reveals a contradiction inherent in holding to the same Hilbert space for both interacting and free fields under the standard assumptions. It states that such fields cannot satisfy the canonical commutation relations consistently if treated as identical to their asymptotic counterparts. This complicates the interaction picture, requiring modified representations of field operators that handle actual interactions separately from their asymptotic forms, necessitating a rigorous non-perturbative approach to field quantization.

5. Elucidate Ruelle's formal framework for handling scattering in QFT, particularly focusing on the limitations presented by Haag's theorem.

    Solution: Ruelle's framework circumvents Haag's theorem by defining scattering based on localized observables and operators, ensuring that asymptotic fields are not merely idealizations of their interacting versions. By focusing on operators within confined spacetime regions, Ruelle allows for the conceptual differentiation of in-states and out-states, thus bypassing the issue that arises in Haag's theorem concerning operator representation, and enforcing continuity through localized measurement processes rather than blanket assumptions.

6. Analyze the significance of Haag-Ruelle theory in understanding non-perturbative phenomena in quantum field theory. How does it underlie the consistent modeling of particle interactions?

    Solution: Haag-Ruelle theory lays the groundwork for describing fully interacting quantum fields non-perturbatively by providing tools for analyzing particle scatterings that concern complexities beyond perturbative regimes. By focusing on operator formalism and asymptotic field definitions, it ensures adherence to physical principles like locality and micro-causality. This methodology underlines accurate modeling of non-trivial phenomena like particle creation and maintenance of causality, crucial when exploring quantum field dynamics beyond simple interaction scenarios.

Practice Problems 2

1. Define the asymptotic fields of a quantum field $\phi(x)$ in terms of the Hamiltonian H. Write down the mathematical expressions for both the incoming and outgoing fields.

2. Explain the significance of asymptotic completeness in scattering theory. How is the Hilbert space $\mathcal{H}$ described in terms of incoming and outgoing Fock spaces?

3. Derive the S-matrix expression given the overlap of the asymptotic fields ϕ_{in} and ϕ_{out}.

4. Discuss Haag's theorem and its implications for canonical quantization in QFT. Under what assumptions does the theorem present challenges?

5. Describe how Ruelle's formulation overcomes the limitations presented by Haag's theorem. What conditions does Ruelle's approach impose on the asymptotic fields?

6. What roles does Haag-Ruelle theory play in the broader context of non-perturbative phenomena in QFT? Provide examples of how it's applied to complex interactions.

Answers 2

1. Define the asymptotic fields of a quantum field $\phi(x)$ in terms of the Hamiltonian H. Write down the mathematical expressions for both the incoming and outgoing fields.

   Solution: Asymptotic fields are defined to capture the free behavior of fields at temporal infinities, providing the basis to study scattering processes. The expressions are:

   For incoming fields:
   $$\phi_{in}(x) = \lim_{t \to -\infty} e^{iHt} \phi(x) e^{-iHt}$$

   For outgoing fields:
   $$\phi_{out}(x) = \lim_{t \to +\infty} e^{iHt} \phi(x) e^{-iHt}$$

   Both forms use the time evolution operator e^{iHt}, corresponding to the Hamiltonian H that describes the energy dynamics of the system.

2. Explain the significance of asymptotic completeness in scattering theory. How is the Hilbert space $\mathcal{H}$ described in terms of incoming and outgoing Fock spaces?

   Solution: Asymptotic completeness asserts that the entire Hilbert space of physical states $\mathcal{H}$ can be decomposed into contributions from both incoming and outgoing free states, denoted by Fock spaces. This is expressed as:

   $$\mathcal{H} = \mathcal{H}_{in} \oplus \mathcal{H}_{out}$$

   This decomposition illustrates that any state observed at scattering limits can be explained using these free (non-interacting) states. This concept is fundamental in guaranteeing that no information about the system's dynamics is missed in the scattering framework.

3. Derive the S-matrix expression given the overlap of the asymptotic fields ϕ_{in} and ϕ_{out}.

Solution: The S-matrix, or scattering matrix, represents the transition amplitude between incoming and outgoing states. The overlap, or inner product, of ϕ_{in} and ϕ_{out} determines these amplitudes:

$$\langle \phi_{out}|S|\phi_{in}\rangle$$

This matrix element provides the probability amplitude for the system to transition from an in-state to an out-state, reflecting the interaction dynamics encoded in the quantum field theory.

4. Discuss Haag's theorem and its implications for canonical quantization in QFT. Under what assumptions does the theorem present challenges?

Solution: Haag's theorem highlights the intricate problem of field quantization in interacting theories. It states that under the assumption of a shared Hilbert space for both free and interacting fields, it may be impossible to maintain appropriate commutation relations simultaneously. Its consequences imply:

- Distinguishing interacting fields from asymptotic fields is necessary.
- Canonical quantization requires careful construction to ensure proper representation.

These challenges occur due to inherent divergences and non-trivial restructuring of field operators within a single Hilbert space framework.

5. Describe how Ruelle's formulation overcomes the limitations presented by Haag's theorem. What conditions does Ruelle's approach impose on the asymptotic fields?

Solution: Ruelle circumvents Haag's theorem's issues through formulating scattering with explicitly localized fields. Key conditions include:

- Utilizing local operators distinctively applicable to asymptotic regions.
- Ensuring limits like:

$$\lim_{t\to\pm\infty} \langle \psi|\phi_{out}(t)\phi_{in}(-t)|\psi\rangle < \infty$$

This ensures finite probabilities for transitions and respects locality and micro-causality principles, essential in physically realistic setups.

6. What roles does Haag-Ruelle theory play in the broader context of non-perturbative phenomena in QFT? Provide examples of how it's applied to complex interactions.

Solution: Haag-Ruelle theory provides tools for the analysis of non-perturbative phenomena where perturbation theory fails. Examples include:

- Analyzing bound states and solitons: Asymptotic field definitions help describe stable, localized states within interacting field models.
- Understanding confinement in gauge theories: Forms foundational work to explore how non-perturbative effects like quark confinement manifest.

It solidifies frameworks that explore high-spin fields, interactions in curved space, and complex external fields, granting insight into real-world quantum behaviors.

Practice Problems 3

1. Consider the asymptotic fields in the Haag-Ruelle theory. Show that if

$$\phi_{in}(x) = \lim_{t\to-\infty} e^{iHt}\phi(x)e^{-iHt}$$

then

$$\frac{d}{dt}e^{iHt}\phi(x)e^{-iHt} = i[H, e^{iHt}\phi(x)e^{-iHt}].$$

4cm

2. Prove that for quantum field $\phi(x)$, the limit

$$\phi_{out}(x) = \lim_{t \to +\infty} e^{iHt}\phi(x)e^{-iHt}$$

implies that

$$\phi_{out}(x) = e^{iHt}\phi(x)e^{-iHt}\big|_{t \to \infty}.$$

4cm

3. Examine the asymptotic completeness of the Hilbert space $\mathcal{H}$ by verifying

$$\mathcal{H} = \mathcal{H}_{in} \oplus \mathcal{H}_{out}$$

through the application of the Fock space decomposition.

4cm

4. Analyze the implications of Haag's theorem by showing that a single Hilbert space $\mathcal{H}$ cannot simultaneously accommodate free and interacting fields with the same operator commutation relations.

4cm

5. Derive the condition for Ruelle's formulation concerning localized observables that governs the finite probability of scattering:

$$\lim_{t \to \pm\infty} \langle \psi | \phi_{out}(t)\phi_{in}(-t) | \psi \rangle < \infty.$$

4cm

6. Using Haag-Ruelle theory, discuss how asymptotic analysis facilitates understanding of non-perturbative phenomena in quantum field theory.

4cm

Answers 3

1. **Solution:** Start by using the property of the Hamiltonian in a quantum field:

$$\frac{d}{dt}\left(e^{iHt}\phi(x)e^{-iHt}\right) = \lim_{h \to 0} \frac{e^{iH(t+h)}\phi(x)e^{-iH(t+h)} - e^{iHt}\phi(x)e^{-iHt}}{h}$$

Recognize the commutator definition:

$$= \lim_{h \to 0} \frac{e^{iHt}(e^{iHh}\phi(x)e^{-iHh} - \phi(x))e^{-iHt}}{h}$$

$$= e^{iHt}\left(\lim_{h \to 0} \frac{e^{iHh}\phi(x)e^{-iHh} - \phi(x)}{h}\right)e^{-iHt}$$

By the definition of operator commutator:

$$= e^{iHt}[iH, \phi(x)]e^{-iHt} = i[H, e^{iHt}\phi(x)e^{-iHt}]$$

Therefore, the derivative confirms inner movement through the commutator.

2. **Solution:** By definition of asymptotic states, we assume:

$$\phi_{out}(x) = \lim_{t \to +\infty} e^{iHt}\phi(x)e^{-iHt}$$

Implying continuity at limits:

$$= \lim_{\epsilon \to 0} e^{iH(t+\epsilon)}\phi(x)e^{-iH(t+\epsilon)}$$

$$= e^{iHt}\phi(x)e^{-iHt}\big|_{t \to \infty}$$

This proves asymptotic field continuity at infinity.

3. **Solution:** Start with total decomposition of states:

$$\mathcal{H}_{in} \oplus \mathcal{H}_{out} = \mathcal{H}$$

This includes basis for (in) and (out) components. For completeness, use projection operators:

$$\mathcal{P}_{in} = \sum |in\rangle\langle in|, \quad \mathcal{P}_{out} = \sum |out\rangle\langle out|$$

These projections cover $\mathcal{H}_{in}$ and $\mathcal{H}_{out}$ fully:

$$\mathcal{I} = \mathcal{P}_{in} + \mathcal{P}_{out}$$

Thus proving any state in $\mathcal{H}$ can be decomposed fully into asymptotic states.

4. **Solution:** According to Haag's theorem, operator sets are:

$$[\phi_{free}(x), \phi_{free}(y)] = [\phi_{int}(x), \phi_{int}(y)]$$

Yet, both fields cannot share identical spatial representations $\mathcal{H}$:

$$\mathcal{H}_{free} \neq \mathcal{H}_{int}$$

due to discrepancies in symmetries and dynamic interactions, indicating creation shifts:

$$\langle 0_{int}|\phi_{int}(x)\phi_{int}(y)|0_{int}\rangle \neq \langle 0_{free}|\phi_{free}(x)\phi_{free}(y)|0_{free}\rangle$$

This proves the field separation requirement per Haag's theorem.

5. **Solution:** Establish elementary limits for localized observables:

$$\lim_{t \to \pm\infty} \langle \psi|\phi_{out}(t)\phi_{in}(-t)|\psi\rangle$$

Observables evaluated become unitary:

$$\phi_{out}(t) = e^{iHt}\phi(x)e^{-iHt}, \quad \phi_{in}(-t) = e^{-iHt}\phi(x)e^{iHt}$$

Evaluation within local framework yields finite results constrained by causality, thus holding finite scattering possibilities.

6. **Solution:** Asymptotic analysis simplifies nonlinear occurrences by addressing large-time behavior:

$$\phi_{in/out} \to \text{Fock space orthogonality}$$

Here, particles elucidate interaction constraints. Non-perturbative fields assessed by:

$$S \approx 1 + iT$$

As computational expansions therein become tractable, focusing on non-trivial regimes like energy states in complex interactions.

Chapter 21

Algebraic Quantum Field Theory

Introduction to Algebraic Methods

Algebraic quantum field theory (AQFT) utilizes abstract algebraic structures to describe quantum fields. Unlike conventional field theory approaches that rely primarily on spacetime functions, AQFT represents observables as elements of a *C**-algebra. The framework seeks to offer a mathematically rigorous foundation, emphasizing the role of operator algebras in capturing quantum field properties and interactions.

C*-Algebras in Quantum Field Theory

A *C**-algebra $\mathcal{A}$ is a Banach algebra with an involution satisfying the *C**-identity:

$$\|a^*a\| = \|a\|^2, \quad \forall a \in \mathcal{A}.$$

This identity is crucial for ensuring the algebraic structure supports a realistic interpretation of quantum mechanics. In AQFT, observable quantities are modeled as self-adjoint elements within a *C**-algebra, which leads to their spectral properties being linked to measurable values.

1 Examples of C*-Algebras

Common examples include:
   - The algebra of bounded operators $\mathcal{B}(\mathcal{H})$ on a Hilbert space $\mathcal{H}$. - Commutative *C**-algebras, which are isomorphic to spaces of continuous functions on a compact Hausdorff space, illustrating the Gelfand-Naimark theorem.

Von Neumann Algebras

Von Neumann algebras, or W*-algebras, extend *C**-algebras, maintaining additional closure properties in the weak operator topology. A von Neumann algebra $\mathcal{M}$ on Hilbert space $\mathcal{H}$ is a *C**-algebra that is closed in the weak operator topology and contains the identity operator.

1 Properties and Examples

Critical features include:
   - Closure under the weak operator topology. - The double commutant theorem: $\mathcal{M} = \mathcal{M}''$.
   Examples encompass:
   - $\mathcal{B}(\mathcal{H})$, the set of all bounded operators on $\mathcal{H}$. - $L^\infty(X, \mu)$, the algebra of essentially bounded measurable functions on a measure space (X, μ).

Applications in Quantum Field Theory

In AQFT, the algebraic framework allows handling quantum fields beyond individual particle interpretations. The association of observable algebras with spacetime regions effectively organizes fields into local entities.

1 Local Algebras and Quantum Fields

For any bounded spacetime region $\mathcal{O}$, a local algebra $\mathcal{A}(\mathcal{O})$ is defined to encapsulate the observables pertinent to $\mathcal{O}$. These algebras should satisfy Haag-Kastler axioms, including isotony and locality:

$$\text{If } \mathcal{O}_1 \subseteq \mathcal{O}_2, \text{ then } \mathcal{A}(\mathcal{O}_1) \subseteq \mathcal{A}(\mathcal{O}_2).$$

$$[\mathcal{A}(\mathcal{O}_1), \mathcal{A}(\mathcal{O}_2)] = 0, \text{ if } \mathcal{O}_1 \cap \mathcal{O}_2 = \emptyset.$$

2 State Spaces and Representations

Given a *C**-algebra $\mathcal{A}$, states can be represented as positive linear functionals $\omega : \mathcal{A} \to \mathbb{C}$ satisfying $\|\omega\| = 1$. Via the GNS (Gelfand-Naimark-Segal) construction, a state ω induces a representation $(\pi, \mathcal{H}_\omega)$, where $\pi : \mathcal{A} \to \mathcal{B}(\mathcal{H}_\omega)$.

The GNS construction highlights:
- Associating $\mathcal{A}$ with operators on Hilbert space $\mathcal{H}_\omega$. - Realizing physical states as vectors in $\mathcal{H}_\omega$.

Implementing Algebraic Methods

Algorithmic approaches in AQFT may utilize:

Algorithm 4: Algebraic Methodology in Quantum Field Theory

Result: Develop algebraic frameworks for quantum fields.
while *require AQFT analysis* **do**
> Define local algebras $\mathcal{A}(\mathcal{O})$ for spacetime regions;
> Ensure $\mathcal{A}(\mathcal{O}_1) \subseteq \mathcal{A}(\mathcal{O}_2)$ under inclusion;
> Validate commutativity for disjoint regions;
> Construct state space representations via GNS;

end

Practice Problems 1

1. Define a *C**-algebra $\mathcal{A}$. Then, prove that for any element $a \in \mathcal{A}$, it holds that $\|a^*\| = \|a\|$.

2. Explain the significance of von Neumann algebras being closed in the weak operator topology. Illustrate this property using $\mathcal{B}(\mathcal{H})$.

3. Discuss the GNS (Gelfand-Naimark-Segal) construction for a *C**-algebra and demonstrate how it leads to a representation of the algebra on a Hilbert space.

4. For a given compact region $\mathcal{O}$ in spacetime, what are the properties that the local algebra $\mathcal{A}(\mathcal{O})$ must satisfy according to the Haag-Kastler axioms?

5. Define the isotony property of local algebras in the context of algebraic quantum field theory. Provide a formal proof for its necessity.

6. Using the double commutant theorem, substantiate why von Neumann algebras can be considered extensions of *C**-algebras.

Answers 1

1. **Define a *C**-algebra $\mathcal{A}$. Then, prove that for any element $a \in \mathcal{A}$, it holds that $\|a^*\| = \|a\|$.**

   Solution: A *C**-algebra $\mathcal{A}$ is a Banach algebra with an involution, such that $\|a^*a\| = \|a\|^2$ for all $a \in \mathcal{A}$.

- Consider the element $a \in \mathcal{A}$. By the *C** identity, we have:

$$\|a^*a\| = \|a\|^2$$

- Replacing a with a^*:

$$\|(a^*)^*a^*\| = \|a^*\|^2$$

- Since $(a^*)^* = a$, the equation becomes:

$$\|aa^*\| = \|a^*\|^2$$

- We conclude that $\|a^*\| = \|a\|$ because:

$$\|aa^*\| = \|a^*a\|$$

completes the commutative property desired.

2. **Explain the significance of von Neumann algebras being closed in the weak operator topology. Illustrate this property using $\mathcal{B}(\mathcal{H})$.**

   Solution:

   - A von Neumann algebra $\mathcal{M}$ is a *C**-algebra that is closed in the weak operator topology, capturing infinite-dimensional phenomena more thoroughly.

   - The importance of closure in the weak operator topology means that limits of sequences of operators within $\mathcal{M}$, when evaluated against suitable vector states, stay within $\mathcal{M}$.

   - In the case of $\mathcal{B}(\mathcal{H})$, weak topology closure ensures the algebra handles convergence in Hilbert space operations robustly without losing integral operators.

3. **Discuss the GNS (Gelfand-Naimark-Segal) construction for a *C**-algebra and demonstrate how it leads to a representation of the algebra on a Hilbert space.**

   Solution:

   - Given a state ω on *C**-algebra $\mathcal{A}$, the GNS construction associates a Hilbert space $\mathcal{H}_\omega$.

   - The process works by forming equivalence classes from the algebra based on kernel elements of ω defining: $\ker \omega = \{a \in \mathcal{A} \mid \omega(a^*a) = 0\}$.

   - These classes, combined via $\langle a + \ker, b + \ker \rangle = \omega(a^*b)$, extend into $\mathcal{H}_\omega$.

   - The algebra acts on $\mathcal{H}_\omega$ by $\pi(a)(b+\ker) = ab+\ker$, completing $(\pi, \mathcal{H}_\omega)$ as a unitary representation.

4. **For a given compact region $\mathcal{O}$ in spacetime, what are the properties that the local algebra $\mathcal{A}(\mathcal{O})$ must satisfy according to the Haag-Kastler axioms?**

   Solution:

   - The local algebra $\mathcal{A}(\mathcal{O})$ satisfies:
   - **Isotony:** If $\mathcal{O}_1 \subseteq \mathcal{O}_2$, then $\mathcal{A}(\mathcal{O}_1) \subseteq \mathcal{A}(\mathcal{O}_2)$.
   - **Locality:** If $\mathcal{O}_1 \cap \mathcal{O}_2 = \emptyset$, then $[\mathcal{A}(\mathcal{O}_1), \mathcal{A}(\mathcal{O}_2)] = 0$.
   - **Covariance:** Dynamics respect symmetry transformations like Poincaré transformations.

5. **Define the isotony property of local algebras in the context of algebraic quantum field theory. Provide a formal proof for its necessity.**

   Solution:

   - **Definition:** The isotony property stipulates that if $\mathcal{O}_1 \subseteq \mathcal{O}_2$, then $\mathcal{A}(\mathcal{O}_1) \subseteq \mathcal{A}(\mathcal{O}_2)$, maintaining algebra hierarchy with spatial containment.

   - **Necessity Proof:**

- Assume the opposite, $\mathcal{A}(\mathcal{O}_1) \not\subseteq \mathcal{A}(\mathcal{O}_2))$. Logically, this breaks spatially dependent logical coherency where more command entails more descriptions.
- Following traditional interpretative rules, $\mathcal{A}(\mathcal{O}_1)$ must inform $\mathcal{A}(\mathcal{O}_2)$, which larger entropic regions obey. Disagreement transgresses basic causality.

6. **Using the double commutant theorem, substantiate why von Neumann algebras can be considered extensions of *C**-algebras.**

Solution:

- The double commutant theorem states: for any *C**-algebra $\mathcal{A}$ that includes identity and is bounded, its weak closure coincides with its second commutant, $\mathcal{A}''$.
- This translates $\mathcal{A}'' = \overline{\mathcal{A}}^{\text{w.o.}}$, establishing that weak closures manifest 'further layered' structure compared to regular operator topology.
- Consequently, von Neumann algebras represent *C**-algebras grown to unpack closure nuances through weak convergence.

Practice Problems 2

1. Define a *C**-algebra and explain the significance of the *C**-identity in the context of quantum mechanics.

2. Illustrate with an example, how a commutative *C**-algebra is related to spaces of continuous functions on a compact Hausdorff space.

3. Discuss the differences between *C**-algebras and von Neumann algebras with respect to their closure properties and their importance in quantum field theory.

4. Explain the Haag-Kastler axioms in the context of local algebras and their role in algebraic quantum field theory.

5. Describe the GNS construction and how it associates a *C**-algebra with operators on a Hilbert space, including the realization of physical states.

6. Analyze the significance of the weak operator topology in von Neumann algebras and its implications for quantum field theory.

Answers 2

1. Define a *C**-algebra and explain the significance of the *C**-identity in the context of quantum mechanics. **Solution:**
   A *C**-algebra $\mathcal{A}$ is a Banach algebra equipped with an involution $a \mapsto a^*$ satisfying the *C**-identity $\|a^*a\| = \|a\|^2$ for all $a \in \mathcal{A}$. This identity is crucial because it ensures that the self-adjoint elements of the algebra can be understood as observables in quantum mechanics, where the spectral properties correspond to measurable quantities.

2. Illustrate with an example, how a commutative *C**-algebra is related to spaces of continuous functions on a compact Hausdorff space. **Solution:**
   A classic example is the algebra $C(X)$, where X is a compact Hausdorff space, consisting of all continuous complex-valued functions on X. According to the Gelfand-Naimark theorem, any commutative *C**-algebra is isometrically isomorphic to the algebra of continuous functions on its Gelfand spectrum, further illustrating the deep interplay between functional analysis and topology.

3. Discuss the differences between *C**-algebras and von Neumann algebras with respect to their closure properties and their importance in quantum field theory. **Solution:**

While *C**-algebras are closed under the norm topology, von Neumann algebras, also known as W*-algebras, are closed in the weak operator topology. This closure under weak topology means that von Neumann algebras can include limits of bounded measurable functions, making them a critical structure for formulating observables and states in quantum field theory, particularly in the presence of infinitely many degrees of freedom.

4. Explain the Haag-Kastler axioms in the context of local algebras and their role in algebraic quantum field theory. **Solution:**
The Haag-Kastler axioms, also known as the axioms of local quantum physics, dictate how local algebras operate in AQFT. These include:
- Isotony: If $\mathcal{O}_1 \subseteq \mathcal{O}_2$, then $\mathcal{A}(\mathcal{O}_1) \subseteq \mathcal{A}(\mathcal{O}_2)$.
- Locality: $[\mathcal{A}(\mathcal{O}_1), \mathcal{A}(\mathcal{O}_2)] = 0$ if $\mathcal{O}_1 \cap \mathcal{O}_2 = \emptyset$. These axioms ensure that the algebraic structures respect the causal and structural aspects of spacetime, crucially encoding the quantum field constraints.

5. Describe the GNS construction and how it associates a *C**-algebra with operators on a Hilbert space, including the realization of physical states. **Solution:**
The GNS (Gelfand-Naimark-Segal) construction is a method to represent *C**-algebras on a Hilbert space. Starting with a state ω, a positive linear functional on a *C**-algebra $\mathcal{A}$, the construction provides a Hilbert space $\mathcal{H}_\omega$, a representation $\pi : \mathcal{A} \to \mathcal{B}(\mathcal{H}_\omega)$, and a cyclic vector ξ_ω such that $\omega(a) = \langle \pi(a)\xi_\omega, \xi_\omega \rangle$. Thus, it paves the way for associating algebraic states with physical states as vectors.

6. Analyze the significance of the weak operator topology in von Neumann algebras and its implications for quantum field theory. **Solution:**
The weak operator topology is significant in von Neumann algebras because it allows them to be closed under limits of sequences of operators with respect to pointwise convergence on the Hilbert space. This property makes them particularly well-suited for describing quantum systems with infinitely many degrees of freedom, as often encountered in QFT. This closure property is essential for ensuring that all physically relevant observables, which may manifest as limits of other observables, are contained within the algebra.

Practice Problems 3

1. Using the concept of *C**-algebras, prove that the algebra of bounded operators $\mathcal{B}(\mathcal{H})$ on a Hilbert space $\mathcal{H}$ is a *C**-algebra.

2. Show that for any bounded spacetime region $\mathcal{O}$, the local algebra $\mathcal{A}(\mathcal{O})$ satisfies the isotony property in algebraic quantum field theory.

3. Consider the von Neumann algebra $L^\infty(X, \mu)$ on a measure space (X, μ). Demonstrate that it is closed under the weak operator topology.

4. Explain the significance of the GNS construction in representing states in algebraic quantum field theory. How does it connect states and physical systems?

5. Verify that the operator commutation relation $[\mathcal{A}(\mathcal{O}_1), \mathcal{A}(\mathcal{O}_2)] = 0$ holds when $\mathcal{O}_1 \cap \mathcal{O}_2 = \emptyset$.

6. Discuss the role of *C**-identity $\|a^*a\| = \|a\|^2$ in ensuring the physical interpretability of observables in quantum mechanics.

Answers 3

1. Using the concept of *C**-algebras, prove that the algebra of bounded operators $\mathcal{B}(\mathcal{H})$ on a Hilbert space $\mathcal{H}$ is a *C**-algebra.

Solution:
A *C**-algebra is a Banach algebra with an involution satisfying the *C**-identity:

$$\|a^*a\| = \|a\|^2, \quad \forall a \in \mathcal{A}.$$

For $\mathcal{B}(\mathcal{H})$:
- Completeness: $\mathcal{B}(\mathcal{H})$ is complete with respect to the operator norm, satisfying the Banach space property.
- Involution: The adjoint operator T^* exists for all $T \in \mathcal{B}(\mathcal{H})$, acting as an involution.
- *C**-identity: The norm $\|T^*T\|$ equals $\|T\|^2$ because T^*T is self-adjoint, ensuring the spectral radius $\sqrt{\|T^*T\|}$ matches $\|T\|$.

Therefore, $\mathcal{B}(\mathcal{H})$ is a *C**-algebra.

2. Show that for any bounded spacetime region $\mathcal{O}$, the local algebra $\mathcal{A}(\mathcal{O})$ satisfies the isotony property in algebraic quantum field theory.

Solution:
The isotony property states:

$$\text{If } \mathcal{O}_1 \subseteq \mathcal{O}_2, \text{ then } \mathcal{A}(\mathcal{O}_1) \subseteq \mathcal{A}(\mathcal{O}_2).$$

.
- Observables in $\mathcal{A}(\mathcal{O}_1)$ naturally map into $\mathcal{A}(\mathcal{O}_2)$ by definition, as $\mathcal{O}_1$ is contained within $\mathcal{O}_2$.
- This inclusion ensures that the algebraic structure respects spatial order and containment.

Therefore, isotony is satisfied.

3. Consider the von Neumann algebra $L^\infty(X, \mu)$ on a measure space (X, μ). Demonstrate that it is closed under the weak operator topology.

Solution:

- $L^\infty(X, \mu)$ is the space of essentially bounded measurable functions and acts as a von Neumann algebra.
- Closure under the weak operator topology involves:
- Convergence: A sequence of functions $f_n \to f$ converges weakly.
- Boundedess is preserved, as the weak limit f respects essential bounds.
- Weak operator topology closure is ensured by the preservation of measurable and bounded properties in weak limits.

Thus, $L^\infty(X, \mu)$ is closed in the weak operator topology.

4. Explain the significance of the GNS construction in representing states in algebraic quantum field theory. How does it connect states and physical systems?

Solution:

- The GNS construction starts from a state ω, a linear functional on a *C**-algebra $\mathcal{A}$.
- It produces a Hilbert space $\mathcal{H}_\omega$ using equivalence classes derived from ω-zero elements.
- A representation $\pi : \mathcal{A} \to \mathcal{B}(\mathcal{H}_\omega)$ is formed, aligning algebra elements with operators.
- Physical states are realized as vectors in $\mathcal{H}_\omega$, linking quantum states to measurable realities.

Thus, GNS provides a bridge between algebraic structures and observable quantum systems.

5. Verify that the operator commutation relation $[\mathcal{A}(\mathcal{O}_1), \mathcal{A}(\mathcal{O}_2)] = 0$ holds when $\mathcal{O}_1 \cap \mathcal{O}_2 = \emptyset$.

Solution:

- The relation $[\mathcal{A}(\mathcal{O}_1), \mathcal{A}(\mathcal{O}_2)] = 0$ emerges from locality and relativistic covariance.
- Observables in disjoint regions commute as no causal influence bridges $\mathcal{O}_1$ and $\mathcal{O}_2$.
- This reflects microcausality in AQFT, ensuring causally independent algebras act simultaneously without interferences.

The given condition guarantees commuting observables.

6. Discuss the role of *C**-identity $\|a^*a\| = \|a\|^2$ in ensuring the physical interpretability of observables in quantum mechanics.

   Solution:

   - The *C**-identity reinforces that norms remain consistent across operations.
   - It implies a connection between algebraic elements and self-adjoint operators representing observables.
   - Self-adjointness ensures real eigenvalues, foundational to quantum mechanical interpretations where observable values must be real numbers.

   This identity is integral to aligning abstract algebra with concrete physical phenomena.

Chapter 22

Operator Product Expansion and Microcausality

Introduction to Operator Product Expansion

Operator product expansion (OPE) is a powerful tool in quantum field theory used to express the product of two operators at nearby points as a series involving local operators. For two operators $\mathcal{O}_1(x)$ and $\mathcal{O}_2(y)$ with x and y close together, the expansion is written as:

$$\mathcal{O}_1(x)\mathcal{O}_2(y) \sim \sum_i C_i(x-y)\mathcal{O}_i\left(\frac{x+y}{2}\right).$$

Here, $C_i(x-y)$ are the `Wilson` coefficients that encapsulate the short-distance behavior, and $\mathcal{O}_i$ are local operators. The choice of operators $\mathcal{O}_i$ is determined by the symmetries and dynamics of the quantum field theory under consideration.

Convergence of the Operator Product Expansion

The series in the operator product expansion is generally not convergent in the traditional sense. Instead, it offers an asymptotic series, capturing the behavior of the operator product in the limit as the separation $|x-y|$ approaches zero. Mathematically, this is expressed as:

$$\lim_{\epsilon \to 0} \left\| \mathcal{O}_1(x+\epsilon)\mathcal{O}_2(x) - \sum_{i=1}^{N} C_i(\epsilon)\mathcal{O}_i(x) \right\| = 0,$$

for some order N and small ϵ. The utility of OPE lies in its ability to simplify complex quantum field interactions by focusing on dominant contributions.

Microcausality in Field Theory

Microcausality is a principle guiding quantum field theories, ensuring that observables separated by spacelike intervals commute. This holds as a fundamental requirement for maintaining causality aligned with the principles of relativity. For operators $\mathcal{A}(x)$ and $\mathcal{B}(y)$, microcausality is represented by:

$$[x-y]^2 < 0 \implies [\mathcal{A}(x), \mathcal{B}(y)] = 0.$$

This equation stresses that no information or influence can travel faster than light, preserving the cause-effect relationship fundamental to physical theories.

1 Implications of Microcausality on OPE

Microcausality imposes constraints on the operator product expansion. Specifically, the non-commuting terms in the expansion must vanish for spacelike separations. Analyzing the dependencies of $\mathcal{O}_1$ and $\mathcal{O}_2$ at such separations leads to:

$$C_i(x - y) = 0 \quad \text{for spacelike separations.}$$

Ensuring microcausality holds means developing OPE expressions that reflect these cancellations accurately.

Algorithmic Approach in OPE Calculation

Calculating the coefficients and constructing the operator product expansion often involves complex computations. An algorithmic approach can streamline these calculations by:

Algorithm 5: Algorithm for OPE Construction
Result: Compute OPE coefficients
while *require precise OPE expansion* **do**
Identify relevant operators $\mathcal{O}_i$ to the field theory;
Calculate short-distance singularities of operator products;
Determine Wilson coefficients $C_i(x - y)$;
Validate convergence and microcausality constraints;
end

Examples of OPE in Quantum Field Theory

Examples in quantum electrodynamics (QED) and quantum chromodynamics (QCD) demonstrate OPE's application:

- **QED:** In electron-photon interactions, the OPE aids in decomposing higher-order loop diagrams into series expansions, focusing on the charged current coupling.

- **QCD:** The expansion elucidates the behavior of quarks and gluons at high energies, pivotal for understanding deep inelastic scattering processes.

These examples highlight OPE's ubiquity and necessity in simplifying quantum interactions grounded in the fundamental fields.

Practice Problems 1

1. Discuss the role of Wilson coefficients $C_i(x - y)$ in the operator product expansion (OPE) and analyze their dependence on short-distance behavior within quantum field theory.

2. Demonstrate the mathematical expression and concept behind the convergence of the OPE, specifying why it is usually considered an asymptotic series rather than a convergent one.

3. Explain microcausality in quantum field theory, elaborating on its significance in maintaining causality. Discuss its implications for the commutation relations of spacelike separated operators.

4. Derive the conditions under which the Wilson coefficients $C_i(x - y)$ must vanish in the context of microcausality, given the relation $[x - y]^2 < 0 \implies [\mathcal{A}(x), \mathcal{B}(y)] = 0$.

5. Develop an algorithmic approach to compute the OPE, detailing each step required for determining the Wilson coefficients and ensuring microcausality is respected.

6. Provide examples of the operator product expansion in quantum electrodynamics (QED) and quantum chromodynamics (QCD), highlighting the importance of OPE in simplifying complex interactions.

Answers 1

1. **Wilson Coefficients in OPE:** Wilson coefficients $C_i(x-y)$ encapsulate the short-distance (ultraviolet) behavior of the operator product $\mathcal{O}_1(x)\mathcal{O}_2(y)$ in the OPE. In the series expansion:

$$\mathcal{O}_1(x)\mathcal{O}_2(y) \sim \sum_i C_i(x-y)\mathcal{O}_i\left(\frac{x+y}{2}\right),$$

the coefficients are determined primarily by the symmetries of the quantum field theory and the scaling dimensions of the operators involved. They encode information about the high-energy or short-distance interactions and are essential in understanding the renormalization and effective field theory framework.

2. **Convergence of OPE:** The OPE is formulated as an asymptotic series rather than a convergent series. This means it approximates the behavior of the operator product as the separation $|x-y|$ approaches zero, rather than summing precisely to the operator product for finite separations. The mathematical representation:

$$\lim_{\epsilon \to 0}\left\|\mathcal{O}_1(x+\epsilon)\mathcal{O}_2(x) - \sum_{i=1}^{N} C_i(\epsilon)\mathcal{O}_i(x)\right\| = 0,$$

implies that the series captures the dominant contributions in the limit of small ϵ, effectively simplifying calculations by focusing on terms that grow slowest as $|x-y|$ diminishes.

3. **Microcausality:** In quantum field theory, microcausality ensures that observables associated with spacelike separated events commute. Given:

$$[x-y]^2 < 0 \implies [\mathcal{A}(x), \mathcal{B}(y)] = 0,$$

the principle safeguards the causal order dictated by relativity, implying no information or influence can propagate faster than light. This restriction helps maintain invariant causal structures across reference frames, making it a cornerstone of relativistic quantum theories.

4. **Vanishment of Wilson Coefficients:** For a scalar quantity like the Wilson coefficient $C_i(x-y)$ to respect microcausality, it must vanish for spacelike separation $[x-y]^2 < 0$. This ensures the vanishing of any contributing terms in the OPE that violate the commutation:

$$[x-y]^2 < 0 \implies C_i(x-y) = 0.$$

The linkage lies in the causality requirement, guiding the expansion to inherently exclude terms violating causality constraints.

5. **Algorithmic Computation of OPE:** To compute OPE with algorithmic clarity, the approach involves:

   - **Identifying Relevant Operators**: Asses the operators $\mathcal{O}_i$ within the theory.
   - **Computing Short-Distance Singularities**: Determine $C_i(x-y)$ by analyzing the behavior near ultraviolet (short distance) limits.
   - **Calculating Wilson Coefficients**: Use symmetry arguments and renormalization conditions to compute $C_i(x-y)$.
   - **Ensuring Microcausality**: Validate that $C_i(x-y)$ respects spacelike separation commutation relations.

   These steps outline a systematic pathway to precise OPE construction.

6. **Examples of OPE:** Within QED and QCD, OPE proves instrumental:

   - **QED**: Involving electron-photon interactions, OPE can deconstruct complex loop diagrams into tractable series form, focusing on primary coupling currents.
   - **QCD**: It elucidates quark-gluon interactions at higher energy regimes, aiding investigations into deep inelastic scattering where OPE effectively computes non-perturbative hadronic effects.

   Thus, OPE stands as a vital conceptual toolkit in theoretical particle physics.

Practice Problems 2

1. Demonstrate the conditions under which the operator product expansion (OPE) series converges to a meaningful description of operator interactions. Analyze the role of asymptotic behavior within your demonstration.

2. Consider two operators $\mathcal{O}_1(x)$ and $\mathcal{O}_2(y)$. If $[x - y]^2 < 0$, explain why microcausality ensures that these operators commute. Use the principles of special relativity in your explanation.

3. Construct an explicit example of OPE in quantum electrodynamics (QED) and explain how Wilson coefficients are determined within this context. Specify any assumptions made during the construction.

4. Given the operator product $\mathcal{O}_1(x)\mathcal{O}_2(x)$, derive the form of the asymptotic expansion as $|x - y| \to 0$ using a hypothetical field theory of your choice. Discuss the selection of operators and their relevance.

5. Discuss how microcausality impacts the application of OPE in quantum chromodynamics (QCD), particularly in terms of high-energy processes and the simplification of theoretical calculations.

6. Design an algorithm to calculate the Wilson coefficients in a generic field theory. Detail the step-by-step approach, emphasizing the computational roles at each stage.

Answers 2

1. **Solution:** The operator product expansion (OPE) is said to converge in the asymptotic sense when the series does not converge traditionally but provides an accurate description of operator behavior as the separation $|x - y|$ approaches zero. Mathematically, as $\epsilon \to 0$,

$$\left\| \mathcal{O}_1(x + \epsilon)\mathcal{O}_2(x) - \sum_{i=1}^{N} C_i(\epsilon)\mathcal{O}_i(x) \right\| \to 0.$$

The OPE captures the leading contributions at short distances, encapsulated by Wilson coefficients $C_i(x - y)$, reflecting the hierarchy of physical interactions decreasing with separation.

2. **Solution:** Microcausality in quantum field theory mandates that for operators $\mathcal{A}(x)$ and $\mathcal{B}(y)$ separated by spacelike intervals $[x - y]^2 < 0$,

$$[\mathcal{A}(x), \mathcal{B}(y)] = 0.$$

This condition preserves causality as dictated by special relativity, ensuring no superluminal information transfer. Observables cannot influence each other outside of the light cone, thus preserving deterministic cause-effect relations.

3. **Solution:** In quantum electrodynamics (QED), consider the product of current operators $j^\mu(x)$,

$$j^\mu(x)j^\nu(y) \sim \sum_i C_i^{\mu\nu}(x - y)\mathcal{O}_i\left(\frac{x + y}{2}\right).$$

The determination of Wilson coefficients $C_i^{\mu\nu}(x - y)$ relies on perturbative expansions and Feynman diagrams analysis. Assumptions include gauge invariance and weak field conditions, focusing on short-distance singularities.

4. **Solution:** In a hypothetical scalar field theory, express $\mathcal{O}_1(x)\mathcal{O}_2(y)$ as:

$$\mathcal{O}_1(x)\mathcal{O}_2(y) \sim \sum_i C_i(x-y)\phi^i\left(\frac{x+y}{2}\right),$$

where ϕ is the fundamental field. The operators ϕ^i are chosen for their relevance to symmetry and dynamics, and the series provides an asymptotic descriptor as $|x-y| \to 0$.

5. **Solution:** In QCD, microcausality affects OPE by delineating non-commuting terms for operators at spacelike separations. This property simplifies theoretical calculations by exploiting degrees of freedom at high energies, particularly in parton distribution functions during deep inelastic scattering.

6. **Solution:** An algorithm to compute Wilson coefficients generally involves:

Algorithm 6: Algorithm for Wilson Coefficient Calculation

Result: Compute Wilson coefficients

while *OPE required* **do**

    Identify operators $\mathcal{O}_i$ relevant to theory;

    Determine short-distance singularities from operator products;

    Calculate coefficients $C_i(x-y)$ via perturbation techniques;

    Verify convergence and validate microcausality;

end

This approach iteratively improves precision and maintains conformity with theoretical constraints.

Practice Problems 3

1. Describe the general form of the operator product expansion (OPE) for two operators $\mathcal{O}_1(x)$ and $\mathcal{O}_2(y)$ in quantum field theory. Include the role of Wilson coefficients and local operators.

2. Analyze the convergence properties of the operator product expansion. Explain whether the OPE is a convergent series or an asymptotic one and justify your answer.

3. Define microcausality in the context of quantum field theory and elucidate its mathematical representation for two operators $\mathcal{A}(x)$ and $\mathcal{B}(y)$.

4. Explain how microcausality influences the form and constraints of the operator product expansion, particularly in terms of the coefficients $C_i(x - y)$.

5. Propose an algorithmic approach for calculating OPE coefficients, identifying the steps involved in ensuring both convergence and microcausality constraints are satisfied.

6. Provide examples of how the operator product expansion is applied in quantum electrodynamics (QED) and quantum chromodynamics (QCD), emphasizing the simplification it offers.

Answers 3

1. **General Form of OPE:**

$$\mathcal{O}_1(x)\mathcal{O}_2(y) \sim \sum_i C_i(x - y)\mathcal{O}_i\left(\frac{x + y}{2}\right).$$

Solution: The operator product expansion is a way to express the product of two operators at nearby points as a series of local operators, $\mathcal{O}_i$, weighted by Wilson coefficients, $C_i(x - y)$. These coefficients encapsulate the short-distance (or high-energy) behaviors between the points x and y. The choice of local operators is dictated by the symmetries and dynamics of the underlying quantum field theory.

2. **Convergence Properties of the OPE:**

The OPE is generally an asymptotic series rather than a convergent one. **Solution:** Mathematically, an asymptotic series in the context of OPE means:

$$\lim_{\epsilon \to 0}\left\| \mathcal{O}_1(x + \epsilon)\mathcal{O}_2(x) - \sum_{i=1}^{N} C_i(\epsilon)\mathcal{O}_i(x) \right\| = 0.$$

This indicates that for any finite order N, the series approximates the operator product well when $|x-y|$ is small, even if the series does not converge in the traditional sense. The utility lies in capturing the dominant contributions in this limit.

3. **Definition of Microcausality:**

For two operators $\mathcal{A}(x)$ and $\mathcal{B}(y)$, microcausality is expressed as:

$$[x-y]^2 < 0 \implies [\mathcal{A}(x), \mathcal{B}(y)] = 0.$$

Solution: Microcausality is a principle ensuring that observables separated by spacelike intervals must commute, maintaining causality in accordance with relativity. It implies that no information or physical influences propagate faster than the speed of light.

4. **Impact of Microcausality on OPE:**

Microcausality imposes constraints such that non-commuting terms in the OPE must vanish for spacelike separations.

Solution: Thus, for the series:

$$C_i(x-y) = 0 \quad \text{for spacelike separations.}$$

This condition ensures that the OPE respects causality, necessitating that expansions reflecting terms would cancel appropriately in such regimes.

5. **Algorithmic Approach for OPE Construction:**

Algorithm 7: Algorithm for OPE Construction

Result: Compute OPE coefficients

while *require precise OPE expansion* **do**

    Identify relevant operators $\mathcal{O}_i$ to the field theory;

    Calculate short-distance singularities of operator products;

    Determine Wilson coefficients $C_i(x-y)$;

    Validate convergence and microcausality constraints;

end

Solution: This approach systematically identifies required structures, evaluates short-distance behaviors, and enforces essential constraints to produce a viable OPE reflecting the theory's dynamics.

6. **Examples of OPE Applications:**

In QED, the OPE is utilized in decomposing electron-photon interaction diagrams.

Solution: It breaks down complex high-order diagrams into manageable series, isolating dominant charged current contributions. In QCD, OPE helps parse quark-gluon dynamics at high energies, crucial for exploring deep inelastic scattering processes. These applications demonstrate how OPE offers simplifying power in complex field-theoretic calculations.

Chapter 23

Renormalization and Functional Analysis

Introduction to Renormalization in Quantum Field Theory

The process of renormalization addresses the infinities that arise in quantum field theories. Renormalization involves re-expressing quantities like mass and charge in terms of physically observable parameters. Consider a perturbative expansion of a field:

$$S[\phi] = \int \left(\frac{1}{2} \partial_\mu \phi \partial^\mu \phi - \frac{m^2}{2} \phi^2 - \frac{\lambda}{4!} \phi^4 \right) d^4 x.$$

Here, λ represents the coupling constant, and divergences in calculations lead to the need for a renormalization procedure.

Regularization Techniques

Regularization serves as an intermediate step where a parameter is introduced to tame infinities. Popular methods include:

1   Dimensional Regularization

This method involves calculating integrals in a spacetime of $d = 4 - \epsilon$ dimensions. The integral takes the form:

$$I(\epsilon) = \int \frac{d^d p}{(2\pi)^d} \frac{1}{p^2 + m^2}.$$

As $\epsilon \to 0$, singularities manifest, which are then managed by introducing counterterms.

2   Pauli-Villars Regularization

In Pauli-Villars regularization, additional fields with large masses M are introduced to cancel out divergences:

$$I_{PV} = \int \frac{d^4 p}{(2\pi)^4} \left(\frac{1}{p^2 + m^2} - \frac{1}{p^2 + M^2} \right).$$

This makes the integral converge due to the subtraction of divergent parts.

Renormalization Group Equations

Renormalization group (RG) equations describe how physical parameters change with variations in the energy scale μ. The renormalization group equation is expressed as:

$$\mu \frac{d}{d\mu} g(\mu) = \beta(g(\mu)),$$

where $g(\mu)$ is the renormalized coupling constant and β is the beta function.

Functional Analysis in Renormalization

Functional analysis offers tools to handle spaces of functions, crucial for the rigorous treatment of quantum fields.

1 Banach and Hilbert Spaces

Quantum fields can be treated as elements of Banach and Hilbert spaces, facilitating the decomposition of interactions. Consider the inner product:

$$\langle f|g \rangle = \int f(x)g(x)\,dx.$$

These spaces provide a structured framework for analyzing operator convergence in renormalization.

2 Operator Theory

Operators in functional spaces represent observables and dynamics in quantum field theories. The spectral theorem allows the decomposition of self-adjoint operators:

$$A = \int \lambda\,dE_\lambda,$$

enabling a deeper understanding of the spectrum of quantum operators relevant to renormalization.

Application of Functional Analysis in Regularization and Renormalization

1 Spectral Theory in Field Regularization

Spectral theory aids in regularizing operators by providing insights into eigenvalue distributions and transformations necessary for practical renormalization.

2 Constructive Field Theory

Constructive field theory employs functional analysis to construct interactions within quantum fields rigorously, ensuring the finite nature of calculated predictions.

Algorithm 8: Renormalization and RG Flow

Result: Renormalization Process
Start with a bare Lagrangian;
Apply regularization method like dimensional regularization;
Introduce counterterms to cancel divergences;
while $\beta(g(\mu)) \neq 0$ **do**
| Use the RG to evolve parameters as a function of the scale;
end
End with physical observables expressed in terms of renormalized parameters;

Practice Problems 1

1. Derive the renormalization group equation using dimensional regularization for a given coupling constant λ in the Lagrangian:

$$\mathcal{L} = \frac{1}{2}(\partial\phi)^2 - \frac{m^2}{2}\phi^2 - \frac{\lambda}{4!}\phi^4.$$

2. Explain the role of counterterms in Pauli-Villars regularization when additional fields are introduced:

$$\mathcal{L}_{PV} = \mathcal{L} + \sum_i c_i \left(\frac{1}{2}(\partial\phi_i)^2 - \frac{M_i^2}{2}\phi_i^2\right).$$

3. Apply functional analysis by determining the conditions under which a self-adjoint operator is defined in a Hilbert space associated with renormalization:

$$\langle f|Ag\rangle = \langle Af|g\rangle.$$

4. Utilize spectral theory to discuss operator regularization and its importance in maintaining a finite spectrum after renormalization.

5. Provide a detailed outline of how constructive field theory applies functional analysis to ensure finite predictions in interacting quantum fields.

6. Explain how the beta function $\beta(g)$ influences the stability of a quantum field theory under energy scale changes.

Answers 1

1. Derivation of the renormalization group equation begins with the dimensional regularization of the Lagrangian:

$$\mathcal{L}_d = \frac{1}{2}(\partial\phi)^2 - \frac{m^2}{2}\phi^2 - \frac{\lambda}{4!}\phi^4,$$

where the spacetime dimensions are extended to $d = 4 - \epsilon$. The dimensionless coupling constant is redefined as:

$$\lambda \to \mu^\epsilon \lambda(\mu),$$

where μ is the renormalization scale. Calculating the divergent diagrams gives:

$$\lambda(\mu) = Z_\lambda \lambda_R(\mu),$$

with Z_λ being the renormalization constant. Differentiating with respect to μ, we obtain the renormalization group equation:

$$\mu\frac{d}{d\mu}\lambda(\mu) = \beta(\lambda(\mu)) = -\epsilon\lambda(\mu) + \text{higher-order terms.}$$

2. In Pauli-Villars regularization, counterterms serve to cancel the divergences that arise due to high-energy behavior by introducing additional massive fields ϕ_i:

$$\mathcal{L}_{PV} = \mathcal{L} + \sum_i c_i \left(\frac{1}{2}(\partial \phi_i)^2 - \frac{M_i^2}{2} \phi_i^2 \right).$$

These fields have coefficients c_i chosen to ensure cancellations occur. As M_i become arbitrarily large, their contributions vanish, leaving a regulated Lagrangian:

$$\tilde{\mathcal{L}} = \mathcal{L} + \mathcal{L}_{PV} \sim \mathcal{L}_{\text{finite}}.$$

3. A self-adjoint operator A in a Hilbert space is one that satisfies:

$$\langle f | Ag \rangle = \langle Af | g \rangle.$$

The domain of A must match with the domain where these equality conditions hold. For renormalization, A represents observables like the Hamiltonian, satisfying:

$$D(A) = \{ f \in \mathcal{H} \mid \langle f | Ag \rangle, \langle Af | g \rangle \text{ are defined} \}.$$

This ensures Hermitian operators are equivalently described by their spectral components.

4. Spectral theory is pivotal in ensuring operator regularization through its decomposition:

$$A = \int \lambda \, dE_\lambda,$$

where dE_λ is the spectral measure. This decomposition allows examination of eigenvalues indicating stability and convergence of the renormalized theory:

$$\sigma(A) \subseteq \text{finite subset of real numbers.}$$

By appropriately choosing regularization, one can maintain finiteness without introducing non-physical eigenvalues.

5. Constructive field theory applies functional analysis to rigorously establish the existence of interactions by ensuring all calculations lead to finite and consistent predictions. This involves formulating:

$$\text{Setting up local algebras for observables,}$$
$$\text{Defining Hamiltonians with proper domain.}$$

Utilizing Banach and Hilbert space structures, every perturbative order remains finite by establishing convergence conditions:

$$\lim_{n \to \infty} \| H_n - H \| = 0, \text{ for dynamics } H.$$

6. The beta function $\beta(g)$ encapsulates the behavior of coupling constants during scale transformations:

$$\mu \frac{d}{d\mu} g(\mu) = \beta(g(\mu)),$$

with stability indicated by $\beta(g_*) = 0$, where g_* is a fixed point. Whether $\beta(g)$ is positive, negative, or zero determines:

$$\text{IR (Infrared) or UV (Ultraviolet) stable points;}$$

these critical points guide the theory's predictive range and consistency across scales, ensuring that physical observables are renormalized without divergences.

Practice Problems 2

1. Consider the Lagrangian for a scalar field theory given by:

$$S[\phi] = \int \left(\frac{1}{2} \partial_\mu \phi \partial^\mu \phi - \frac{m^2}{2} \phi^2 - \frac{\lambda}{4!} \phi^4 \right) d^4x.$$

   Derive the equation of motion for the scalar field ϕ using the Euler-Lagrange equation.

2. For the integration involved in dimensional regularization:

$$I(\epsilon) = \int \frac{d^d p}{(2\pi)^d} \frac{1}{p^2 + m^2},$$

   perform the integration and express the result in terms of the gamma function Γ and ϵ.

3. Consider the Pauli-Villars regularization:

$$I_{PV} = \int \frac{d^4 p}{(2\pi)^4} \left(\frac{1}{p^2 + m^2} - \frac{1}{p^2 + M^2} \right).$$

   Explain how this regularization cancels the infinities in loop integrals.

4. Using the renormalization group equation:

$$\mu \frac{d}{d\mu} g(\mu) = \beta(g(\mu)),$$

   find the fixed points g^* when $\beta(g^*) = 0$, and discuss their stability.

5. Demonstrate how functional analysis, particularly Hilbert spaces, is applied in the decomposition of quantum field interactions.

6. Utilize the spectral theorem in the context of renormalization to describe the decomposition of self-adjoint operators.

Answers 2

1. **Equation of Motion for a Scalar Field:**

$$S[\phi] = \int \left(\frac{1}{2} \partial_\mu \phi \partial^\mu \phi - \frac{m^2}{2} \phi^2 - \frac{\lambda}{4!} \phi^4 \right) d^4 x.$$

Solution: The Euler-Lagrange equation is given by:

$$\frac{\partial \mathcal{L}}{\partial \phi} - \partial_\mu \left(\frac{\partial \mathcal{L}}{\partial (\partial_\mu \phi)} \right) = 0.$$

Calculate $\frac{\partial \mathcal{L}}{\partial \phi}$:

$$\frac{\partial \mathcal{L}}{\partial \phi} = -m^2 \phi - \frac{\lambda}{3!} \phi^3.$$

Calculate $\frac{\partial \mathcal{L}}{\partial (\partial_\mu \phi)}$:

$$\frac{\partial \mathcal{L}}{\partial (\partial_\mu \phi)} = \partial^\mu \phi.$$

Apply the Euler-Lagrange equation:

$$-m^2 \phi - \frac{\lambda}{3!} \phi^3 - \partial_\mu (\partial^\mu \phi) = 0 \implies \Box \phi + m^2 \phi + \frac{\lambda}{6} \phi^3 = 0.$$

Therefore, the equation of motion is:

$$\Box \phi + m^2 \phi + \frac{\lambda}{6} \phi^3 = 0.$$

2. **Integrating using Dimensional Regularization:**

$$I(\epsilon) = \int \frac{d^d p}{(2\pi)^d} \frac{1}{p^2 + m^2}.$$

Solution: Convert to spherical coordinates and evaluate the integral:

$$I(\epsilon) = K_d \int_0^\infty \frac{p^{d-1}\, dp}{p^2 + m^2}.$$

Use the substitution $x = p^2/m^2$:

$$I(\epsilon) = K_d \int_0^\infty \frac{(mx)^{(d/2)-1} m^{1/2}\, dx}{1 + x}.$$

Relate to Γ:

$$I(\epsilon) \propto \Gamma\left(\frac{d/2 - 1}{2}\right)\Gamma\left(1 - \frac{d}{2}\right).$$

Thus, expressing in ϵ:

$$I(\epsilon) \approx \frac{1}{\epsilon} + \Gamma - function \text{ terms}.$$

3. **Pauli-Villars Regularization:**

$$I_{PV} = \int \frac{d^4 p}{(2\pi)^4}\left(\frac{1}{p^2 + m^2} - \frac{1}{p^2 + M^2}\right).$$

Solution: The subtraction term $\frac{1}{p^2 + M^2}$ introduces a high mass scale M, which when subtracted from the low mass scale integral helps to negate the divergent terms as $M \to \infty$. Therefore:

$$\lim_{M \to \infty} I_{PV} \rightsquigarrow \text{converges, achieving regularization.}$$

4. **Fixed Points of RG Equation:**

$$\mu\frac{d}{d\mu}g(\mu) = \beta(g(\mu)).$$

Solution: At the fixed point: $\beta(g^*) = 0$. Determine stability:

$$\frac{d\beta}{dg}\Big|_{g=g^*} > 0 \implies \text{unstable.}$$

$$\frac{d\beta}{dg}\Big|_{g=g^*} < 0 \implies \text{stable.}$$

Thus the stability depends on the sign of the derivative of the beta function at g^*.

5. **Hilbert Spaces in QFT: Solution:** Use Hilbert spaces for quantum state representation:

$$\mathcal{H} \ni \phi(x) \sim |\phi\rangle.$$

Project operator:

$$A = \int \lambda\, dE_\lambda \implies A|\phi\rangle = H|\phi\rangle,$$

analyzing interaction through eigenstates.

6. **Spectral Theorem in Renormalization: Solution:** For self-adjoint operators:

$$A = \int \lambda\, dE_\lambda,$$

Implement in renormalization:

$$Use\ spectral\ decomposition\ to\ analyze\ eigenvalue\ shifts\ under\ renormalization.$$

Practice Problems 3

1. Evaluate the following integral using dimensional regularization:

$$I(\epsilon) = \int \frac{d^d p}{(2\pi)^d} \frac{1}{p^2 + m^2}$$

2. Show how the Pauli-Villars regularization technique is applied to evaluate:

$$I_{PV} = \int \frac{d^4 p}{(2\pi)^4} \left(\frac{1}{p^2 + m^2} - \frac{1}{p^2 + M^2} \right)$$

3. Demonstrate the calculation of the beta function $\beta(g)$ from the renormalization group equation:

$$\mu \frac{d}{d\mu} g(\mu) = \beta(g(\mu))$$

4. Verify the spectral decomposition of a self-adjoint operator A in terms of the spectral measure:

$$A = \int \lambda \, dE_\lambda$$

5. Explain the concept of weak convergence in Hilbert spaces and provide an example involving quantum fields.

6. Outline the steps to construct a renormalized field theory using the tools of functional analysis.

Answers 3

1. **Solution:**

   Begin with the definition of dimensional regularization where the integral is evaluated in $d = 4 - \epsilon$.

   $$I(\epsilon) = \int \frac{d^d p}{(2\pi)^d} \frac{1}{p^2 + m^2}$$

   Use the formula for the integral in d-dimensions:

   $$I(\epsilon) = \frac{\pi^{d/2}}{(2\pi)^d} \Gamma(1 - \frac{d}{2})(m^2)^{\frac{d}{2} - 1}$$

   Substitute $d = 4 - \epsilon$:

   $$I(\epsilon) = \frac{\pi^{(4-\epsilon)/2}}{(2\pi)^{4-\epsilon}} \Gamma(\frac{\epsilon}{2})(m^2)^{-\epsilon/2}$$

   Analytically continue to $\epsilon = 0$ to isolate the pole.

   The result shows how the singularity appears as $\epsilon \to 0$.

2. **Solution:**

   Apply the Pauli-Villars regularization:

   $$I_{PV} = \int \frac{d^4 p}{(2\pi)^4} \left(\frac{1}{p^2 + m^2} - \frac{1}{p^2 + M^2} \right)$$

   For large M, the subtraction cancels the UV divergence.

   Simplify the integral:

$$I_{PV} = \int \frac{d^4p}{(2\pi)^4} \frac{M^2 - m^2}{(p^2 + m^2)(p^2 + M^2)}$$

Evaluating, the divergence in $\frac{1}{p^2+m^2}$ is canceled out by the subtraction term.

This regularized result is finite as $M \to \infty$.

3. **Solution:**

The RG equation is:

$$\mu \frac{d}{d\mu} g(\mu) = \beta(g(\mu))$$

Assume $g(\mu) = g_0 + c \ln(\frac{\mu}{\mu_0})$.

Take the derivative:

$$\frac{d}{d\mu}(g_0 + c \ln(\frac{\mu}{\mu_0})) = \frac{c}{\mu}$$

So:

$$\mu \frac{d}{d\mu} g(\mu) = c = \beta$$

This confirms β as the constant governing the change of $g(\mu)$ with scale.

4. **Solution:**

For a self-adjoint operator A:

$$A = \int \lambda \, dE_\lambda$$

Establish E_λ as the projection-valued measure.

Verify properties:

- E_λ is non-decreasing: $\lambda_1 \leq \lambda_2 \Rightarrow E_{\lambda_1} \leq E_{\lambda_2}$.

- If $\{E_\lambda\}$ is complete, then the spectral projector resolves to the identity when integrated over the spectrum.

Confirm through orthogonality and ranges that decomposition holds for any vector.

5. **Solution:**

Weak convergence in Hilbert spaces involves sequences $\{f_n\}$ such that:

$$\langle f_n, g \rangle \to \langle f, g \rangle$$

Provide example: sequence of wavefunctions $\psi_n(x)$ converges weakly to $\psi(x)$ based on expectation values.

This is crucial for studying quantum states' convergence properties.

6. **Solution:**

Outline the steps as follows:

- Begin with the bare Lagrangian of the theory.

- Apply regularization techniques to each divergent integral.

- Introduce counterterms to the Lagrangian to absorb infinite parts.

- Utilize the machinery of Banach/Hilbert spaces to ensure field convergence.

- Employ operator theory to manage the transformed operators and establish renormalization conditions.

These functional analysis tools help assure self-consistency and correctness of renormalized predictions.

Chapter 24

Functional Integrals and Path Integrals

Introduction to Functional Integration

Functional integration extends the concept of integrating over finite-dimensional spaces to infinite-dimensional spaces, pivotal in quantum field theory and statistical mechanics. Consider a real-valued function $S[x(t)]$ defined over a space of paths $x(t)$, also referred to as a functional. In a finite-dimensional analogy, functional integrals can be represented as:

$$\int \mathcal{D}[x(t)]\, e^{-S[x(t)]},$$

where $\mathcal{D}[x(t)]$ denotes the measure over the space of functions $x(t)$.

Path Integral Formulation of Quantum Mechanics

The path integral approach provides an alternative formulation of quantum mechanics, introduced by Richard Feynman. For a quantum mechanical system described by a Hamiltonian H, the probability amplitude for a transition from an initial state $|q_i, t_i\rangle$ to a final state $|q_f, t_f\rangle$ is given by:

$$\langle q_f, t_f | q_i, t_i \rangle = \int \mathcal{D}[q(t)]\, e^{\frac{i}{\hbar} S[q(t)]},$$

where $S[q(t)] = \int_{t_i}^{t_f} L(q, \dot{q}, t)\, dt$ is the action functional obtained from the Lagrangian $L(q, \dot{q}, t)$.

1   Derivation from Schrödinger's Equation

The path integral formulation can be derived starting from the Schrödinger equation. Consider the time evolution operator:

$$U(t_f, t_i) = e^{-iH(t_f - t_i)/\hbar}.$$

For small time intervals Δt, approximate $U(\Delta t) \approx e^{-iH\Delta t/\hbar}$. Inserting a complete set of position states, the expression becomes:

$$\langle q_{n+1} | U(\Delta t) | q_n \rangle \approx e^{i\Delta t \frac{(q_{n+1} - q_n)^2}{2m\hbar}},$$

leading to an iterative process where:

$$\lim_{N \to \infty} \int \prod_{j=1}^{N-1} dq_j \, \exp\left\{ \frac{i}{\hbar} \Delta t \sum_{j=1}^{N} L(q_j, \dot{q}_j) \right\},$$

with $\dot{q}_j = (q_{j+1} - q_j)/\Delta t$.

Path Integral Formulation in Quantum Field Theory

In quantum field theory, the path integral formalism extends to fields, where one considers:

$$\int \mathcal{D}[\phi(x)] \, e^{iS[\phi]},$$

with $S[\phi]$ being the action for the field $\phi(x)$. The partition function $Z[J]$ in the presence of an external source $J(x)$ is defined as:

$$Z[J] = \int \mathcal{D}[\phi] \, e^{i\left(S[\phi] + \int J(x)\phi(x)\, d^4x \right)}.$$

1 Generating Functionals and Correlation Functions

The generating functional for correlation functions in quantum field theory is derived from the path integral. It is expressed by:

$$Z[J] = \langle 0 | \, e^{i \int J(x)\phi(x)\, d^4x} \, | 0 \rangle.$$

Correlation functions are obtained by functional differentiation with respect to $J(x)$:

$$\langle \phi(x_1)\phi(x_2)\ldots\phi(x_n) \rangle = \frac{1}{Z[0]} \frac{\delta}{\delta J(x_1)} \frac{\delta}{\delta J(x_2)} \cdots \frac{\delta}{\delta J(x_n)} Z[J] \Big|_{J=0}.$$

Feynman Path Integrals: A Computational Approach

Feynman path integrals provide a computational tool for calculating quantum amplitudes through the sum-over-histories approach. Express the propagator as:

$$G(x_f, t_f; x_i, t_i) = \int \mathcal{D}[x(t)] \, e^{\frac{i}{\hbar} S[x(t)]},$$

where the paths $x(t)$ contribute to the transition amplitude between initial and final states.

Algorithm 9: Computational Procedure for Path Integrals

Start with a discrete approximation of the path integral;
Introduce a time slicing with small intervals;
foreach *time slice* Δt **do**
$\quad$ Sum over all intermediate positions;
$\quad$ Compute the contribution of each path;
end
Take the limit as the number of slices goes to infinity;

Functional Methods in Statistical Field Theory

In statistical mechanics, the partition function $\mathcal{Z}$ of a system is computed using functional integrals. For a field theory at finite temperature, the partition function is formulated as:

$$\mathcal{Z} = \int \mathcal{D}[\phi]\, e^{-\frac{1}{\hbar} \int_0^{\beta\hbar} d\tau \int d^3x\, \mathcal{L}(\phi, \partial_\mu \phi)},$$

where $\beta = 1/k_B T$ is the inverse temperature.

1   Connection to Classical Statistical Mechanics

Classical statistical mechanics describes systems using the partition function Z which can be connected to the path integral formalism by identifying:

$$Z = \int \mathcal{D}[\phi]\, e^{-\beta H[\phi]},$$

where $H[\phi]$ resembles the Hamiltonian, bridging quantum and classical descriptions in the functional framework.

Practice Problems 1

1. Derive the expression for the probability amplitude in the path integral formulation of quantum mechanics for a particle traversing a potential $V(q)$ between times t_i and t_f.

$$\int \mathcal{D}[q(t)]\, e^{\frac{i}{\hbar} \int_{t_i}^{t_f} (L(q, \dot{q}, t) - V(q))\, dt}$$

2. Show how the generating functional $Z[J]$ for correlation functions is related to the quantum field theory path integral with an external source $J(x)$.

$$Z[J] = \int \mathcal{D}[\phi]\, e^{i(S[\phi] + \int J(x)\phi(x)\, d^4x)}$$

3. Derive the expression for the partition function $\mathcal{Z}$ using functional integrals at finite temperature.

$$\mathcal{Z} = \int \mathcal{D}[\phi] \, e^{-\frac{1}{\hbar} \int_0^{\beta\hbar} d\tau \int d^3x \, \mathcal{L}(\phi, \partial_\mu \phi)}$$

4. Utilize Feynman's sum-over-histories approach to express the propagator for a free particle.

$$G(x_f, t_f; x_i, t_i) = \int \mathcal{D}[x(t)] \, e^{\frac{i}{\hbar} S[x(t)]}$$

5. Demonstrate the connection between the classical partition function and the path integral formulation.

$$Z = \int \mathcal{D}[\phi] \, e^{-\beta H[\phi]}$$

6. Explain how the path integral formulation is used to derive the uncertainty principle in quantum mechanics.

Answers 1

1. Derive the expression for the probability amplitude in the path integral formulation of quantum mechanics.

   Solution:
   Starting with the definition of the action $S[q(t)] = \int_{t_i}^{t_f} L(q, \dot{q}, t)\, dt$ where L is the Lagrangian, we incorporate the potential $V(q)$ into the Lagrangian, $L = T - (V + V(q))$. Therefore, the adjusted action becomes:

   $$S'[q(t)] = \int_{t_i}^{t_f} \left(L(q, \dot{q}, t) - V(q) \right) dt$$

   The probability amplitude is calculated by:

   $$\langle q_f, t_f | q_i, t_i \rangle = \int \mathcal{D}[q(t)]\, e^{\frac{i}{\hbar} S'[q(t)]}$$

   We substitute $S'[q(t)]$ with our new action, leading to:

   $$\langle q_f, t_f | q_i, t_i \rangle = \int \mathcal{D}[q(t)]\, e^{\frac{i}{\hbar} \int_{t_i}^{t_f} (L(q, \dot{q}, t) - V(q))\, dt}$$

2. Show how the generating functional $Z[J]$ is related to the path integral.

   Solution:
   In quantum field theory, the generating functional is given by:

   $$Z[J] = \langle 0 | e^{i \int J(x)\phi(x)\, d^4 x} | 0 \rangle$$

   This replaces the field operator with the classical field configuration, using path integrals:

   $$Z[J] = \int \mathcal{D}[\phi]\, e^{i(S[\phi] + \int J(x)\phi(x)\, d^4 x)}$$

   The path integral sums over all possible field configurations, incorporating the influence of the source $J(x)$.

3. Derive the expression for the partition function $\mathcal{Z}$ using functional integrals at finite temperature.

   Solution:
   In statistical field theory, the partition function at a finite temperature is defined as:

   $$\mathcal{Z} = \text{Tr}\left(e^{-\beta H} \right)$$

   By representing it in terms of path integrals, it can be rewritten:

   $$\mathcal{Z} = \int \mathcal{D}[\phi]\, e^{-\frac{1}{\hbar} \int_0^{\beta \hbar} d\tau \int d^3 x\, \mathcal{L}(\phi, \partial_\mu \phi)}$$

   This expression arises by substituting the quantum Hamiltonian path integral form into its statistical counterpart.

4. Utilize Feynman's sum-over-histories approach to express the propagator.

   Solution:
   Using Feynman's approach for a free particle:

   $$G(x_f, t_f; x_i, t_i) = \int \mathcal{D}[x(t)]\, e^{\frac{i}{\hbar} S[x(t)]}$$

Here, $S[x(t)]$ is the action for the particle:

$$S[x(t)] = \int_{t_i}^{t_f} \left(\frac{1}{2} m \left(\frac{dx}{dt} \right)^2 \right) dt$$

The successive path integral sums over all possible paths $x(t)$, providing the propagator between two spacetime points.

5. Demonstrate the connection between the classical partition function and the path integral formulation.

   Solution:
   In classical statistical mechanics, the partition function is given by:

   $$Z = \int e^{-\beta H(p,q)} \, dp \, dq$$

   In the path integral framework:

   $$Z = \int \mathcal{D}[\phi] \, e^{-\beta H[\phi]}$$

   This equivalence arises when interpreting $\mathcal{D}[\phi]$ as generalized phase space measure.

6. Explain how the path integral formulation is used to derive the uncertainty principle.

   Solution:
   Path integrals inherently involve all possible paths, linking position and momentum to define uncertainty. Specifically, for two states $|\psi_1\rangle$ and $|\psi_2\rangle$, and non-commutative operators:

   $$[\hat{x}, \hat{p}] = i\hbar$$

   The propagator embodies wavefunction spreading over multiple trajectories:

   $$\langle q_f, t_f | q_i, t_i \rangle \approx \exp \left(-\frac{(x_f - x_i)^2}{2\sigma^2(t)} \right)$$

   Leading to uncertainty:

   $$\Delta x \Delta p \geq \frac{\hbar}{2}$$

   This result stems directly from the formulation's inclusion of path variations.

Practice Problems 2

1. Derive the expression for the transition amplitude in the path integral formulation of quantum mechanics:

   $$\langle q_f, t_f | q_i, t_i \rangle = \int \mathcal{D}[q(t)] \, e^{\frac{i}{\hbar} S[q(t)]}$$

2. Show how the path integral formulation incorporates the classical action $S[q(t)]$ through the Lagrangian L:

$$S[q(t)] = \int_{t_i}^{t_f} L(q, \dot{q}, t)\, dt$$

3. Explain how to obtain correlation functions from the generating functional in quantum field theory:

$$\langle \phi(x_1)\phi(x_2)\ldots\phi(x_n) \rangle = \frac{1}{Z[0]} \frac{\delta}{\delta J(x_1)} \frac{\delta}{\delta J(x_2)} \cdots \frac{\delta}{\delta J(x_n)} Z[J] \bigg|_{J=0}$$

4. Compute the partition function in the path integral formulation for a field theory at finite temperature:

$$\mathcal{Z} = \int \mathcal{D}[\phi]\, e^{-\frac{1}{\hbar} \int_0^{\beta\hbar} d\tau \int d^3x\, \mathcal{L}(\phi, \partial_\mu \phi)}$$

5. Verify the connection between the functional integral and classical statistical mechanics:

$$Z = \int \mathcal{D}[\phi]\, e^{-\beta H[\phi]}$$

6. Describe the computational procedure for path integrals using Feynman's sum-over-histories method.

Answers 2

1. Derive the expression for the transition amplitude in the path integral formulation of quantum mechanics:

$$\langle q_f, t_f | q_i, t_i \rangle = \int \mathcal{D}[q(t)] \, e^{\frac{i}{\hbar} S[q(t)]}$$

Solution: The transition amplitude can be derived by discretizing the time into small intervals, and inserting complete sets of position eigenstates at each interval. For each segment, the transition involves the exponential of the classical action. Summing over all possible paths $q(t)$ results in the path integral:

$$\lim_{N \to \infty} \int \prod_{j=1}^{N-1} dq_j \, \exp\left(\frac{i}{\hbar} \sum_{j=1}^{N} L(q_j, \dot{q}_j) \Delta t \right)$$

As $N \to \infty$, this expression becomes the path integral describing quantum mechanics.

2. Show how the path integral formulation incorporates the classical action $S[q(t)]$ through the Lagrangian L:

$$S[q(t)] = \int_{t_i}^{t_f} L(q, \dot{q}, t) \, dt$$

Solution: The classical action S captures the dynamics of a system, integrating the Lagrangian over time. Discretizing the time interval, the action integral approximates:

$$\sum_{j=1}^{N} L(q_j, \dot{q}_j) \Delta t$$

This form directly links to the Lagrangian used in classical mechanics, and in path integrals, contributes to the phase factor, leading to the quantum amplitudes.

3. Explain how to obtain correlation functions from the generating functional in quantum field theory:

$$\langle \phi(x_1) \phi(x_2) \ldots \phi(x_n) \rangle = \frac{1}{Z[0]} \frac{\delta}{\delta J(x_1)} \frac{\delta}{\delta J(x_2)} \cdots \frac{\delta}{\delta J(x_n)} Z[J] \bigg|_{J=0}$$

Solution: The generating functional $Z[J]$ acts as a tool for deriving correlation functions. By functionally differentiating $Z[J]$ with respect to the source $J(x)$, one singles out field configurations, thus calculating n-point correlation functions. Each differentiation introduces a field at position x_i.

4. Compute the partition function in the path integral formulation for a field theory at finite temperature:

$$\mathcal{Z} = \int \mathcal{D}[\phi] \, e^{-\frac{1}{\hbar} \int_0^{\beta \hbar} d\tau \int d^3 x \, \mathcal{L}(\phi, \partial_\mu \phi)}$$

Solution: Field theory at finite temperature approaches systems by compactifying time in imaginary units, considering a Euclidean space where:

$$t \to -i\tau; \quad \tau \in [0, \beta\hbar]$$

Transitions from quantum to statistical mechanics reformulate the partition function using Euclidean actions, evaluating over the interval $[0, \beta\hbar]$ with thermal periodicity.

5. Verify the connection between the functional integral and classical statistical mechanics:

$$Z = \int \mathcal{D}[\phi] \, e^{-\beta H[\phi]}$$

Solution: Bridging quantum field theory and classical statistical mechanics involves expressing the partition function using functional integrals. Here the Hamiltonian $H[\phi]$ parallels the action in Feynman's formulation, and the factor $e^{-\beta H}$ evokes energy weighting in classical realms.

6. Describe the computational procedure for path integrals using Feynman's sum-over-histories method.
   Solution: Begin by discretizing paths over time intervals. For each interval, assess possible increments in position, computing corresponding contributions $e^{\frac{i}{\hbar} S[x(t)]}$. Each path contributes to quantum amplitudes, with the integration over histories translating to the total energy spectrum:

$$\int \mathcal{D}[x(t)] \, e^{\frac{i}{\hbar} S[x(t)]}$$

For each path, accumulate contributions up to the limit of infinite slices, refining towards continuous evaluation in essence reflecting Feynman's intuitive picture of quantum mechanics.

Practice Problems 3

1. Evaluate the path integral for a free particle in quantum mechanics, where the action $S[x(t)] = \int_0^T \frac{1}{2} m\dot{x}^2 \, dt$. Determine the path integral expression and the propagator $G(x_f, T; x_i, 0)$.

2. Show that the path integral for a simple harmonic oscillator in quantum mechanics can be written in terms of the classical action. Given $S_{cl}[x(t)] = \int_0^T \left(\frac{1}{2} m\dot{x}^2 - \frac{1}{2} m\omega^2 x^2 \right) dt$, express the propagator $G(x_f, T; x_i, 0)$.

3. Derive the generating functional $Z[J]$ for a scalar field theory with the Lagrangian $\mathcal{L} = \frac{1}{2}(\partial_\mu \phi \partial^\mu \phi - m^2 \phi^2)$ and show how to obtain the two-point correlation function.

4. Consider the Euclidean path integral for a quantum field with action
   $S_E[\phi] = \int d^4 x \left(\frac{1}{2}(\nabla \phi)^2 + \frac{1}{2} m^2 \phi^2 + \frac{\lambda}{4!} \phi^4 \right)$. Explain the role of the Euclidean action and determine the expression for the partition function $\mathcal{Z}$.

5. Illustrate the process of functional differentiation by computing
   $\frac{\delta}{\delta J(x')} Z[J]$, given that $Z[J] = \int \mathcal{D}\phi \, e^{i(S[\phi] + \int J(x)\phi(x) \, d^4 x)}$. Explain each step of the derivation.

6. Given the generating functional for a free scalar field theory, $Z_0[J] = \exp\left(-\frac{1}{2} \int d^4 x \, d^4 y \, J(x) \Delta(x - y) J(y) \right)$, find the response function $\Delta(x - y)$ by examining the Gaussian form of $Z_0[J]$.

Answers 3

1. Evaluate the path integral for a free particle in quantum mechanics, where the action $S[x(t)] = \int_0^T \frac{1}{2}m\dot{x}^2\,dt$. Determine the path integral expression and the propagator $G(x_f, T; x_i, 0)$.

   Solution: For a free particle, the action is $S[x(t)] = \int_0^T \frac{1}{2}m\dot{x}^2\,dt$. The solution follows from path integration over all paths, with the functional integral being Gaussian:

   $$G(x_f, T; x_i, 0) = \int \mathcal{D}[x(t)]\, e^{\frac{i}{\hbar}S[x(t)]}.$$

   The Gaussian nature of the action allows us to directly compute this integral:

   $$G(x_f, T; x_i, 0) = \sqrt{\frac{m}{2\pi i \hbar T}}\,\exp\left(\frac{i}{\hbar}\frac{m(x_f - x_i)^2}{2T}\right).$$

   This formulation follows by recognizing the quadratic form in the Gaussian measure and performing the path integration exactly.

2. Show that the path integral for a simple harmonic oscillator in quantum mechanics can be written in terms of the classical action. Given $S_{cl}[x(t)] = \int_0^T \left(\frac{1}{2}m\dot{x}^2 - \frac{1}{2}m\omega^2 x^2\right)dt$, express the propagator $G(x_f, T; x_i, 0)$.

   Solution: For the harmonic oscillator, we evaluate the classical action $S_{cl}[x(t)]$. Solving the classical equations, we have:

   $$S_{cl}[x(t)] = \frac{m\omega}{2\sin(\omega T)}\left[(x_f^2 + x_i^2)\cos(\omega T) - 2x_f x_i\right].$$

   The propagator becomes:

   $$G(x_f, T; x_i, 0) = \sqrt{\frac{m\omega}{2\pi i \hbar \sin(\omega T)}}\,\exp\left(\frac{i}{\hbar}S_{cl}[x(t)]\right).$$

   The derivation involves evaluating the path integral, exploiting the symmetry and periodic boundary conditions inherent in the harmonic oscillator.

3. Derive the generating functional $Z[J]$ for a scalar field theory with the Lagrangian $\mathcal{L} = \frac{1}{2}(\partial_\mu \phi \partial^\mu \phi - m^2\phi^2)$ and show how to obtain the two-point correlation function.

   Solution: The generating functional for a scalar field is defined by:

   $$Z[J] = \int \mathcal{D}\phi\, e^{i\int d^4x(\mathcal{L}+J\phi)}.$$

   Substituting the Lagrangian, we have:

   $$Z[J] = \int \mathcal{D}\phi\, e^{i\int d^4x\left(\frac{1}{2}\phi(\Box + m^2)\phi + J\phi\right)}.$$

   Completing the square in the exponent and using properties of Gaussian integrals, we find:

   $$Z[J] = e^{-\frac{1}{2}\int d^4x\, d^4y\, J(x)\Delta(x-y)J(y)}.$$

   The two-point correlation function is:

   $$\langle \phi(x)\phi(y)\rangle = \frac{\delta}{\delta J(x)}\frac{\delta}{\delta J(y)}Z[J]\Big|_{J=0} = \Delta(x-y),$$

   where $\Delta(x-y)$ is the Green's function for the Klein-Gordon operator.

4. Consider the Euclidean path integral for a quantum field with action
$S_E[\phi] = \int d^4x \left(\frac{1}{2}(\nabla\phi)^2 + \frac{1}{2}m^2\phi^2 + \frac{\lambda}{4!}\phi^4 \right)$. Explain the role of the Euclidean action and determine the expression for the partition function $\mathcal{Z}$.

Solution: The Euclidean action S_E arises from the Wick rotation $t \to -i\tau$, translating the quantum amplitude calculation into a statistical partition function problem. The partition function is given by:

$$\mathcal{Z} = \int \mathcal{D}\phi\, e^{-S_E[\phi]}.$$

The focus here is transitioning from Minkowski to Euclidean space to exploit analytic properties and make the path integral converge, framing it as a thermodynamic partition function for statistical analysis.

5. Illustrate the process of functional differentiation by computing
$\frac{\delta}{\delta J(x')}Z[J]$, given that $Z[J] = \int \mathcal{D}\phi\, e^{i(S[\phi] + \int J(x)\phi(x)\, d^4x)}$. Explain each step of the derivation.

Solution: The functional derivative $\frac{\delta}{\delta J(x')}$ acts on $Z[J]$ as follows:

$$\frac{\delta}{\delta J(x')}e^{i\int d^4x\, J(x)\phi(x)} = i\phi(x')e^{i\int d^4x\, J(x)\phi(x)}.$$

Hence,

$$\frac{\delta Z[J]}{\delta J(x')} = i\int \mathcal{D}\phi\, \phi(x')e^{iS[\phi] + i\int d^4x\, J(x)\phi(x)}.$$

Therefore, evaluating at $J = 0$, the procedure extracts observables from the generating functional, linking source and response functions.

6. Given the generating functional for a free scalar field theory, $Z_0[J] = \exp\left(-\frac{1}{2}\int d^4x\, d^4y\, J(x)\Delta(x-y)J(y)\right)$, find the response function $\Delta(x-y)$ by examining the Gaussian form of $Z_0[J]$.

Solution: The generating functional $Z_0[J]$ is characterized as a Gaussian in J; differentiating:

$$\Delta(x-y) = \frac{\delta^2}{\delta J(x)\delta J(y)}\ln Z_0[J].$$

Taking the derivatives,

$$\Delta(x-y) = -\frac{\delta^2}{\delta J(x)\delta J(y)}\left(\frac{1}{2}\int d^4x\, d^4y\, J(x')\Delta(x'-y')J(y')\right).$$

Simplifying,

$$\Delta(x-y) = \Delta(x-y).$$

This demonstrates how Δ connects fields to their classical counterparts in the path-integral approach, encapsulating dynamics of free-field Green's functions.

Chapter 25

Gauge Theories and BRST Symmetry

Gauge Invariance

A key principle in the construction of modern quantum field theories is gauge invariance. Gauge theories describe fields that remain unchanged under local transformations. The simplest example is electromagnetism, described by the gauge group $U(1)$. For a field $\psi(x)$ transforming as $\psi(x) \to e^{i\alpha(x)}\psi(x)$, the electromagnetic vector potential $A_\mu(x)$ must transform as:

$$A_\mu(x) \to A_\mu(x) + \frac{1}{e}\partial_\mu \alpha(x),$$

to maintain invariance of the Euler-Lagrange equations. The Lagrangian for such a theory is:

$$\mathcal{L} = \bar{\psi}(i\gamma^\mu D_\mu - m)\psi - \frac{1}{4}F_{\mu\nu}F^{\mu\nu},$$

where $D_\mu = \partial_\mu + ieA_\mu$ is the covariant derivative ensuring the gauge invariance, and $F_{\mu\nu} = \partial_\mu A_\nu - \partial_\nu A_\mu$ represents the electromagnetic field tensor.

Quantization of Gauge Fields

Quantization in gauge theories requires careful handling of gauge degrees of freedom. The path integral formulation is utilized, where the generating functional Z in a gauge theory is given by:

$$Z = \int \mathcal{D}A \, \mathcal{D}\psi \, \mathcal{D}\bar{\psi} \, e^{iS[A,\psi,\bar{\psi}]},$$

where $S[A, \psi, \bar{\psi}]$ is the action functional. The gauge fixing condition $G[A] = 0$ is introduced to tackle redundancies, leading to the Faddeev-Popov procedure which modifies the path integral:

$$Z = \int \mathcal{D}A \, \mathcal{D}\psi \, \mathcal{D}\bar{\psi} \, \delta(G[A]) \, \det\left(\frac{\delta G}{\delta \alpha}\right) e^{iS[A,\psi,\bar{\psi}]}.$$

This alteration introduces ghost fields $c, \bar{c}$ ensuring unitarity and renormalizability, leading to the modified action:

$$S_{\text{eff}}[A, \psi, \bar{\psi}, c, \bar{c}] = S[A, \psi, \bar{\psi}] + \int d^4x \, \bar{c}\frac{\delta G}{\delta \alpha}c.$$

BRST Formalism

BRST symmetry provides a powerful framework for the quantization of gauge theories. In BRST formalism, gauge transformations are interpreted as symmetry transformations parameterized by anticommuting parameters. The nilpotent BRST transformation s acts on fields Φ as:

$$s\Phi = \delta_\epsilon \Phi - \delta_{\epsilon\bar{\epsilon}} \Phi.$$

For gauge fields A_μ, the transformation is:

$$sA_\mu = D_\mu c,$$

where D_μ is the covariant derivative. For ghost fields, the transformations are expressed as:

$$sc = -\frac{1}{2} g c \times c, \quad s\bar{c} = b,$$

where b is an auxiliary field introduced to linearize the gauge fixing term and $\times$ denotes the Lie algebra commutator.

1 BRST Charge and Cohomology

The BRST charge Q_{BRST} generates the BRST transformations and satisfies $Q_{BRST}^2 = 0$, implying nilpotency. Physical states $|\text{phys}\rangle$ in the quantum theory are those with:

$$Q_{BRST}|\text{phys}\rangle = 0 \quad \text{and} \quad \langle\text{phys}|Q_{BRST} = 0.$$

Such conditions ensure that physical states are gauge invariant in a cohomological sense, described by the physical Hilbert space being the cohomology of Q_{BRST} at zero eigenvalue.

2 Applications and Implications

BRST symmetry offers a coherent methodology to tackle gauge invariance anomalies. It is crucial for demonstrating renormalizability and unitarity. The BRST invariance allows the systematic treatment of gauge theories at both classical and quantum levels, providing a robust avenue for exploring non-Abelian gauge theories such as Quantum Chromodynamics (QCD).

Algorithm for Gauge Fixing and BRST Quantization

Quantizing a non-Abelian gauge theory requires a consistent implementation of gauge fixing and BRST methods, as detailed in the algorithm below:

Algorithm 10: Quantization of Gauge Fields using BRST Formalism

---
Select a gauge fixing condition $G[A] = 0$;
Compute the Faddeev-Popov determinant $\det\left(\frac{\delta G}{\delta \alpha}\right)$;
Introduce ghost fields $c, \bar{c}$ and an auxiliary field b;
Construct the total action $S_{\text{eff}} = S + S_{\text{ghost}}$;
Verify BRST invariance $sS_{\text{eff}} = 0$;
Utilize BRST charge to ensure physical state conditions in the Hilbert space;

Examples of Gauge Theories

Explorations of gauge theories embody rich structures in the Standard Model of particle physics. Each gauge theory highlights specific invariances: $U(1)$ for electromagnetism, $SU(2)$ for weak interaction, and $SU(3)$ for strong interaction. These gauge symmetries are fundamental in describing forces within the framework of quantum fields.

Practice Problems 1

1. Demonstrate the invariance of the Lagrangian under a $U(1)$ gauge transformation for the electromagnetic field.

2. Show how the Faddeev-Popov procedure modifies the path integral for a given gauge theory with a specific gauge fixing condition.

3. Explain the nilpotent nature of the BRST charge and its implications for physical states in gauge theories.

4. Calculate the BRST transformation for a simple gauge field and show how it maintains gauge invariance.

5. Derive the expression for ghost action in a gauge theory using the Faddeev-Popov determinant.

6. Discuss the role of the BRST symmetry in ensuring renormalizability and unitarity in non-Abelian gauge theories.

Answers 1

1. Demonstrate the invariance of the Lagrangian under a $U(1)$ gauge transformation for the electromagnetic field.

   Solution: The Lagrangian for electromagnetism is given by:

   $$\mathcal{L} = \bar{\psi}(i\gamma^\mu D_\mu - m)\psi - \frac{1}{4}F_{\mu\nu}F^{\mu\nu}$$

   Under the $U(1)$ gauge transformation $\psi(x) \to e^{i\alpha(x)}\psi(x)$ and $A_\mu(x) \to A_\mu(x) + \frac{1}{e}\partial_\mu\alpha(x)$, the covariant derivative transforms as $D_\mu\psi \to e^{i\alpha(x)}(D_\mu\psi)$.

   Substitute into the Lagrangian, observe:

   $$\bar{\psi}(i\gamma^\mu D_\mu - m)\psi \to \bar{\psi}e^{-i\alpha(x)}(i\gamma^\mu D_\mu - m)e^{i\alpha(x)}\psi = \bar{\psi}(i\gamma^\mu D_\mu - m)\psi$$

   The gauge invariance is evident for the kinetic term, and since the field tensor $F_{\mu\nu}$ is gauge invariant, the entire Lagrangian remains unchanged. Hence, the gauge invariance holds.

2. Show how the Faddeev-Popov procedure modifies the path integral for a given gauge theory with a specific gauge fixing condition.

   Solution: Start with the generating functional for a gauge theory:

   $$Z = \int \mathcal{D}A\,\mathcal{D}\psi\,\mathcal{D}\bar{\psi}\,e^{iS[A,\psi,\bar{\psi}]}$$

   Introduce a gauge fixing condition $G[A] = 0$ to ensure uniqueness of gauge configurations. Implement this with a delta function:

   $$Z = \int \mathcal{D}A\,\mathcal{D}\psi\,\mathcal{D}\bar{\psi}\,\delta(G[A])\,e^{iS[A,\psi,\bar{\psi}]}$$

   The Faddeev-Popov determinant $\det\left(\frac{\delta G}{\delta\alpha}\right)$ accounts for redundancy in the gauge choice:

   $$Z = \int \mathcal{D}A\,\mathcal{D}\psi\,\mathcal{D}\bar{\psi}\,\delta(G[A])\,\det\left(\frac{\delta G}{\delta\alpha}\right)e^{iS[A,\psi,\bar{\psi}]}$$

   Insert ghost fields $c, \bar{c}$:

   $$S_{\text{eff}} = S + \int d^4x\,\bar{c}\frac{\delta G}{\delta\alpha}c$$

   The path integral includes ghost field contributions, maintaining covariance and aiding in renormalization.

3. Explain the nilpotent nature of the BRST charge and its implications for physical states in gauge theories.

   Solution: The BRST transformation is generated by a charge Q_{BRST}, satisfying:

   $$Q_{BRST}^2 = 0$$

   Nilpotence implies the transformation of an already transformed field returns zero, intrinsic to symmetry redundancy reduction. Physical states satisfy:

   $$Q_{BRST}|\text{phys}\rangle = 0$$

   The symmetry ensures gauge invariance in the cohomological sense. As $Q_{BRST}^2 = 0$, transitions by Q_{BRST} do not impact physical states, essential in defining consistent quantum field variables under gauge redundancies.

4. Calculate the BRST transformation for a simple gauge field and show how it maintains gauge invariance.

   Solution: For a gauge field A_μ, the BRST transformation given by:

   $$sA_\mu = D_\mu c$$

   Where D_μ is the covariant derivative, verifies gauge transformations are symmetries expressed through ghost fields c. Show invariance:

   The Lagrangian invariant under $A_\mu \to A_\mu + D_\mu c$, it shifts uniquely without altering physical observables, confirmed by construction using ghost terms.

5. Derive the expression for ghost action in a gauge theory using the Faddeev-Popov determinant.

   Solution: The Faddeev-Popov determinant in the path integral is managed by ghost fields:

   $$\det\left(\frac{\delta G}{\delta \alpha}\right)$$

   Evaluate determinant via Grassmann variables $c, \bar{c}$:

   $$\delta(G[A])\det\left(\frac{\delta G}{\delta \alpha}\right) \to \int \mathcal{D}c\,\mathcal{D}\bar{c}\, e^{i\int d^4x\, \bar{c}\frac{\delta G}{\delta\alpha}c}$$

   Result in additional ghost term for action, ensuring the modified Lagrangian maintains gauge symmetry under BRST transformation, suitable for non-Abelian gauge theories.

6. Discuss the role of the BRST symmetry in ensuring renormalizability and unitarity in non-Abelian gauge theories.

   Solution: BRST symmetry reinterprets gauge invariance using a global symmetry approach, crucial for consistent field theory description. It:

   - Allows **renormalizability** by maintaining manifestly gauge-invariant terms in covariant path integral quantization.
   - Ensures **unitarity** since it confines physical states to BRST cohomology, consistent with gauge fixing and ghost compensations.
   - Ensures all gauge redundant variables are systematically eliminated, maintaining physical interpretation coherence.

   Hence, BRST formalism robustly underpins theoretical constructs across varied gauge frameworks.

Practice Problems 2

1. Explain the necessity of gauge fixing in the process of quantizing gauge fields and describe the role of the Faddeev-Popov determinant.

2. Derive the transformation rules for the ghost fields c and $\bar{c}$ under the BRST symmetry, and explain their significance.

3. Explain how the BRST charge Q_{BRST} is constructed and demonstrate how it ensures the physical state conditions in a gauge theory.

4. Discuss the concept of BRST cohomology and its importance in identifying physical states in quantum field theories.

5. Describe the procedure for quantizing a non-Abelian gauge theory using the BRST formalism, including the introduction and role of auxiliary fields.

6. Provide an example of a gauge theory and elucidate how gauge invariance is preserved within it, utilizing BRST symmetry and ghost fields.

Answers 2

1. **Necessity of Gauge Fixing and Faddeev-Popov Determinant:**

   Solution:

   Gauge fixing is necessary to handle the redundant gauge degrees of freedom in the quantization of gauge theories. Without gauge fixing, the path integral formulation would involve an overcounting of physically equivalent gauge configurations, leading to divergences.

   The Faddeev-Popov determinant is introduced to compensate for the gauge fixing condition in the path integral. It ensures the correct weighting of gauge configurations and is derived from the Jacobian of the gauge fixing condition:

   $$\delta(G[A]) \det \left(\frac{\delta G}{\delta \alpha} \right),$$

   where $G[A] = 0$ is the gauge fixing condition and α parameterizes gauge transformations. This determinant is expressed as an integral over ghost fields, which cancel unphysical contributions in loop calculations.

2. **Transformation Rules for Ghost Fields under BRST:**

   Solution:

   Under BRST symmetry, the ghost fields c and $\bar{c}$ have the following transformations:

   $$sc = -\frac{1}{2} gc \times c, \quad s\bar{c} = b,$$

   where b is an auxiliary field introduced to simplify the gauge fixing term, and $\times$ denotes the Lie algebra commutator.

   These transformations maintain the nilpotency property $s^2 = 0$, essential for ensuring the consistency of the quantum gauge theory.

3. **Construction of the BRST Charge and Physical State Conditions:**

   Solution:

   The BRST charge Q_{BRST} generates BRST transformations and is constructed to obey the nilpotency condition $Q_{\text{BRST}}^2 = 0$. It acts on field configurations such that:

   $$Q_{\text{BRST}}|\text{phys}\rangle = 0,$$

   enforcing physical state conditions in the quantum theory. These conditions ensure gauge invariance is preserved at the level of observable quantities in the theory's Hilbert space.

4. **BRST Cohomology and Physical States:**

Solution:

BRST cohomology provides a framework for identifying physical states. In this context, the physical Hilbert space is defined as the cohomology of the BRST charge Q_{BRST} at zero eigenvalue, meaning:

$$\{|\text{phys}\rangle\} = \ker(Q_{\text{BRST}})/\text{im}(Q_{\text{BRST}}).$$

This framework guarantees that only gauge-invariant states contribute to physical predictions.

5. **Quantization Procedure in BRST Formalism:**

Solution:

The procedure involves:

- Choosing a gauge fixing condition $G[A] = 0$.
- Computing the Faddeev-Popov determinant to insert into the path integral.
- Introducing ghost fields $c, \bar{c}$ and an auxiliary field b.
- Formulating the modified action including ghost contributions S_{eff}.
- Checking BRST invariance of the action: $sS_{\text{eff}} = 0$.
- Using the BRST charge to enforce physical state conditions.

6. **Example of a Gauge Theory - Electromagnetism with $U(1)$ Gauge Group:**

Solution:

In electromagnetism, the gauge field A_μ transforms under $U(1)$ as:

$$A_\mu(x) \to A_\mu(x) + \frac{1}{e}\partial_\mu \alpha(x).$$

Gauge invariance is preserved by ensuring the Lagrangian remains unchanged under transformations. BRST symmetry introduces ghost fields that aid in defining a consistent framework. The ghost contributions cancel unphysical longitudinal polarizations of the photon. This ensures the theory's renormalizability and unitarity, crucial for making empirical predictions consistent with experiments.

Practice Problems 3

1. Explain the significance of gauge invariance in the context of electromagnetism, particularly focusing on the transformation properties of the electromagnetic vector potential $A_\mu(x)$.

2. Derive the form of the covariant derivative D_μ that ensures gauge invariance of the Lagrangian in a $U(1)$ gauge theory and explain why it is necessary.

3. Outline the steps involved in the Faddeev-Popov procedure, emphasizing how ghost fields are introduced and why they are needed for path integral quantization of gauge theories.

4. Describe the BRST transformation rules for the gauge field A_μ and the ghost fields c and $\bar{c}$. Why is the nilpotency of the BRST operator crucial in quantum gauge theories?

5. Discuss the role of the BRST charge Q_{BRST} in determining the physical state space in quantum gauge theories. How does it help define the cohomology that represents physical states?

6. Provide a detailed explanation of how gauge fixing and the introduction of ghost fields ensure the renormalizability and unitarity of non-Abelian gauge theories within the BRST framework.

Answers 3

1. **Solution:** Gauge invariance is a pivotal concept ensuring the physical consistency of theories like electromagnetism. For the field $\psi(x)$ which transforms as $\psi(x) \to e^{i\alpha(x)}\psi(x)$, the vector potential $A_\mu(x)$ must transform as:

$$A_\mu(x) \to A_\mu(x) + \frac{1}{e}\partial_\mu \alpha(x),$$

in order to leave the physics described by the Euler-Lagrange equations invariant. This ensures that the Lagrangian remains unchanged under local transformations, making it a fundamental requirement for symmetry and resultant conserved quantities according to Noether's theorem.

2. **Solution:** The covariant derivative D_μ in a $U(1)$ gauge theory is essential to maintaining gauge invariance. Defined as:

$$D_\mu = \partial_\mu + ieA_\mu,$$

it corrects the derivative term to compensate for local $U(1)$ gauge transformations of the field. This adjustment maintains the invariance of the kinetic term of the Lagrangian under gauge transformations, thus ensuring the entire theory respects the symmetry dictated by the gauge group.

3. **Solution:** The Faddeev-Popov procedure involves:

1. Introducing a gauge-fixing condition $G[A] = 0$ to handle gauge redundancies. 2. The path integral is modified by:

$$Z = \int \mathcal{D}A\, \mathcal{D}\psi\, \mathcal{D}\bar{\psi}\, \delta(G[A]) \det\left(\frac{\delta G}{\delta \alpha}\right) e^{iS[A,\psi,\bar{\psi}]}.$$

3. Ghost fields c and $\bar{c}$ are introduced to encapsulate the effect of the determinant term. Their introduction ensures that the theory remains unitary and renormalizable by counteracting the effects of gauge fixing on the path integral measure.

4. **Solution:** The BRST transformations are defined by:

$$sA_\mu = D_\mu c,$$

$$sc = -\frac{1}{2}gc \times c, \quad s\bar{c} = b.$$

The nilpotency $s^2 = 0$ is crucial as it ensures the consistency and closure of BRST transformations, guaranteeing that subsequent transformations do not affect the gauge. Nilpotency is vital for defining a residual symmetry that remains after gauge fixing and is essential for the cohomological interpretation of the physical state space.

5. **Solution:** The BRST charge Q_{BRST} generates BRST transformations:

$$Q_{BRST}|\text{phys}\rangle = 0.$$

It is nilpotent ($Q_{BRST}^2 = 0$), implying that the physical state space is defined as the kernel of Q_{BRST} modulo its image (cohomology). This criterion characterizes the set of gauge invariant states, ensuring they are independent of pure gauge degrees of freedom suppressed by gauge fixing.

6. **Solution:** Renormalizability and unitarity in gauge theories are aided by:

- Gauge fixing imposes conditions that eliminate redundant degrees of freedom.
- Ghost fields c and $\bar{c}$ are instrumental in maintaining unitarity, especially in loop calculations, by cancelling non-physical longitudinal or temporal gauge field contributions.
- The BRST framework rigorously formulates gauge fixing, ensuring consistent incorporation of gauge invariance at every perturbative order, vital for the success of non-Abelian theories like QCD.

Chapter 26

Index Theorems and Applications in QFT

Introduction to Index Theorems

The Atiyah-Singer Index Theorem is a significant result in differential geometry and mathematical physics, connecting the analytical properties of elliptic differential operators on manifolds with topological invariants. For a given elliptic differential operator D defined on a compact manifold M, the theorem provides a relationship between the analytic index (`ind_a`) and the topological index (`ind_t`) of D:

$$\texttt{ind\_a}(D) = \texttt{ind\_t}(D).$$

The analytic index is represented as the difference:

$$\texttt{ind\_a}(D) = \dim(\ker(D)) - \dim(\mathrm{coker}(D)),$$

where $\ker(D)$ denotes the kernel of the operator D, and $\mathrm{coker}(D)$ is the cokernel, typically defined as $\mathrm{coker}(D) = \ker(D^*)$.

Proof Outline of the Atiyah-Singer Index Theorem

The proof of the Atiyah-Singer Index Theorem is extensive and comprises several major mathematical ideas and constructions. A simplified outline involves the following steps:

1 Elliptic Complexes and Symbols

Consider an elliptic differential operator $D : E \to F$, acting between two vector bundles E and F over a manifold M. The operator is characterized by its principal symbol, $\sigma(D)$.

The principal symbol $\sigma(D)$ of D is an essential component in determining the ellipticity of the operator. For each cotangent vector $\xi \in T^*M$, the symbol is a homomorphism between the fibers of E and F:

$$\sigma(D)(x,\xi) : E_x \to F_x.$$

The operator D is elliptic if $\sigma(D)(x,\xi)$ is invertible for all $x \in M$ and non-zero ξ.

2 Topological K-Theory

Topological K-theory provides a framework to define the topological index of an elliptic operator. The K-theory group $K^0(M)$ of the manifold M classifies vector bundles over M. The topological index `ind_t`(D) is defined using the class of the symbol $[\sigma(D)] \in K^0(T^*M)$ and a map to integer values.

3   Heat Equation Approach

The heat equation method, introduced by Atiyah and Bott, connects analysis and topology by considering the evolution of heat kernels. The heat operator e^{-tD^*D} helps to approximate the index:

$$\texttt{ind\_a}(D) = \lim_{t \to 0} \int_M \text{tr}(e^{-tD^*D})(x)\,dx.$$

This integral computes the difference of dimensions of $\ker(D)$ and $\ker(D^*)$ by analyzing the behavior as t approaches 0.

Applications to Anomalies in Quantum Field Theory

The Atiyah-Singer Index Theorem finds applications in quantum field theory, particularly in the study of anomalies.

1   Chiral Anomalies

Chiral anomalies arise in theories with fermionic fields where classical symmetries fail upon quantization. The index theorem aids in computing the anomaly present in the Dirac operator coupled to a gauge field:

$$D = \gamma^\mu(\partial_\mu + iA_\mu),$$

where γ^μ are the gamma matrices and A_μ represents the gauge potential.

The computation involves the calculation of:

$$\texttt{ind\_a}(D) = n_+ - n_-,$$

denoting the difference in the number of positive (n_+) and negative (n_-) chirality zero modes of D.

2   Gravitational Anomalies

In the presence of gravitational fields, the index theorem also provides insight into gravitational anomalies. One considers the Dirac operator associated with the Levi-Civita connection on a curved spacetime manifold.

The anomalies are related to the failure of the classical conservation laws in quantum settings, reflecting topological properties of the underlying spacetime and expressed using index theorems.

Algorithmic Approach to Index Calculation

Algorithm 11: Index Calculation of Elliptic Operators

Data: Elliptic operator D on manifold M
Result: Index of D
Compute the principal symbol $\sigma(D)$;
Check ellipticity: verify if $\sigma(D)(x, \xi)$ is invertible for non-zero ξ;
Determine the K-theory class $[\sigma(D)]$ in $K^0(T^*M)$;
Use the heat equation method to evaluate $\texttt{ind\_a}(D) = \lim_{t \to 0} \int_M \text{tr}(e^{-tD^*D})(x)\,dx$;
Compare $\texttt{ind\_a(D)}$ with topological data to infer $\texttt{ind\_t(D)}$;

Practice Problems 1

1. Explain the significance of the Atiyah-Singer Index Theorem in the context of quantum field theory.

2. Define the analytic index of an elliptic differential operator and provide an example of its computation.

3. Discuss the role of the principal symbol in determining the ellipticity of an operator.

4. Describe how topological K-theory is utilized in formulating the topological index of an elliptic operator.

5. Outline the steps involved in the heat equation approach to compute the analytic index of an operator.

6. Discuss the relevance of the index theorem in identifying chiral anomalies in quantum field theories involving gauge fields.

Answers 1

1. Explain the significance of the Atiyah-Singer Index Theorem in the context of quantum field theory.

   Solution: The Atiyah-Singer Index Theorem connects the analytical and topological properties of differential operators on manifolds. In quantum field theory, it helps quantify anomalies, particularly in contexts where symmetries manifest in classical but not quantum systems. The theorem's ability to relate zero modes of operators to topological features helps in understanding field quantization effects and symmetry-breaking mechanisms.

2. Define the analytic index of an elliptic differential operator and provide an example of its computation.

   Solution: The analytic index ($\mathtt{ind\_a}(D)$) of an elliptic differential operator D is given by:

   $$\mathtt{ind\_a}(D) = \dim(\ker(D)) - \dim(\mathrm{coker}(D)).$$

   For instance, consider an elliptic operator with 3 zero modes (ker) and 1 cokernel element (coker). Thus,
   $$\mathtt{ind\_a}(D) = 3 - 1 = 2.$$

3. Discuss the role of the principal symbol in determining the ellipticity of an operator.

   Solution: The principal symbol $\sigma(D)$ of an operator D is defined for cotangent vectors ξ. For D to be elliptic, $\sigma(D)(x, \xi)$ must be invertible for all $x \in M$ and non-zero ξ. This requirement ensures that the operator does not degenerate under Fourier transformation and solutions maintain regularity, establishing ellipticity criteria.

4. Describe how topological K-theory is utilized in formulating the topological index of an elliptic operator.

   Solution: Topological K-theory categorizes vector bundles using groups like $K^0(M)$ and associates these with the symbol $[\sigma(D)]$ of an operator. The topological index ($\mathtt{ind\_t}(D)$) leverages properties of these groups, linking D's principal symbol to manifold topology. This facilitates a topological representation of the operator's analytical index.

5. Outline the steps involved in the heat equation approach to compute the analytic index of an operator.

   Solution: The heat equation approach involves several steps:

   (a) Determine the operator D and its adjoint D^*.

   (b) Construct the heat operator e^{-tD^*D}.

   (c) Compute the heat kernel trace $\mathrm{tr}(e^{-tD^*D})(x)$.

   (d) Integrate this trace over M and evaluate

   $$\mathtt{ind\_a}(D) = \lim_{t \to 0} \int_M \mathrm{tr}(e^{-tD^*D})(x)\,dx.$$

   (e) Analyze the limit as t approaches zero for index determination.

6. Discuss the relevance of the index theorem in identifying chiral anomalies in quantum field theories involving gauge fields.

   Solution: Chiral anomalies emerge when quantizing fermionic fields interacting with gauge fields, undermining classical symmetry. The index theorem provides a framework for calculating anomalies via the operator index:
   $$D = \gamma^\mu(\partial_\mu + iA_\mu).$$

   The difference between positive and negative chirality zero-modes ($n_+ - n_-$) mirrors this index and consequently reveals the quantifiable expression of the anomaly. The theorem highlights the unavoidable anomaly manifestation tied to manifold topology and gauge field configurations.

Practice Problems 2

1. Consider an elliptic operator D on a manifold M. Describe how you would verify the invertibility of its principal symbol $\sigma(D)$.

2. Given the Dirac operator coupled to a gauge field $D = \gamma^\mu(\partial_\mu + iA_\mu)$, compute the analytic index using the concept of chirality zero modes.

3. Explain how topological K-theory is used to determine the topological index of an elliptic operator.

4. Describe the heat equation approach to calculating the analytic index of an elliptic operator.

5. Demonstrate how the Atiyah-Singer Index Theorem applies to analyzing gravitational anomalies.

6. Outline an algorithmic approach for computing the index of an elliptic operator D using both analytic and topological methods.

Answers 2

1. Consider an elliptic operator D on a manifold M. Describe how you would verify the invertibility of its principal symbol $\sigma(D)$.

   Solution: To verify the invertibility of the principal symbol $\sigma(D)$, we need to:

   (a) Identify the vector bundles E and F over the manifold M such that $D : E \to F$.

   (b) Define the principal symbol $\sigma(D)$ as a function:

   $$\sigma(D)(x, \xi) : E_x \to F_x$$

   for each cotangent vector $\xi \in T^*M$.

   (c) Check that $\sigma(D)(x, \xi)$ is invertible for all $x \in M$ and non-zero ξ. This involves determining that the linear map from E_x to F_x is bijective.

2. Given the Dirac operator coupled to a gauge field $D = \gamma^\mu(\partial_\mu + iA_\mu)$, compute the analytic index using the concept of chirality zero modes.

   Solution: The analytic index of the Dirac operator D is given by the difference in the number of positive and negative chirality zero modes:

   $$\texttt{ind\_a}(D) = n_+ - n_-$$

   To compute this:

   (a) Identify zero modes of the Dirac operator D where $D\psi = 0$.

   (b) Classify these zero modes according to their chirality: n_+ for positive chirality and n_- for negative chirality.

   (c) Subtract the number of negative chirality modes from the number of positive chirality modes to find $\texttt{ind\_a}(D)$.

3. Explain how topological K-theory is used to determine the topological index of an elliptic operator.

   Solution: Topological K-theory determines the topological index by:

   (a) Using the principal symbol $\sigma(D)$ to associate the elliptic operator D with an element in K-theory:

   $$[\sigma(D)] \in K^0(T^*M)$$

   (b) Mapping this K-theory class to an integer using a fixed ring homomorphism known as the topological index.

(c) The topological index, `ind_t`(D) is then defined by this integer value, representing a topological invariant of the elliptic operator.

4. Describe the heat equation approach to calculating the analytic index of an elliptic operator.

Solution: The heat equation method involves:

(a) Considering the heat operator e^{-tD^*D}, which describes the propagation of heat as time approaches zero.

(b) Calculating the index using the trace:

$$\texttt{ind\_a}(D) = \lim_{t \to 0} \int_M \mathrm{tr}(e^{-tD^*D})(x)\,dx$$

(c) This integral determines the difference in the dimensions of the kernels of D and D^* by analyzing the small-time behavior of the heat kernel.

5. Demonstrate how the Atiyah-Singer Index Theorem applies to analyzing gravitational anomalies.

Solution: In analyzing gravitational anomalies, the Atiyah-Singer Index Theorem:

(a) Considers a Dirac operator D on a curved spacetime manifold.

(b) Evaluates the index `ind_a`(D), which indicates the presence of anomalies.

(c) Relates the failure of classical conservation laws in quantum fields to the topology of the underlying manifold through the equation:
$$\texttt{ind\_a}(D) = \texttt{ind\_t}(D)$$

(d) The mismatch (anomaly) between analytical properties (e.g., zero modes of the Dirac operator) and topological insights (from manifold invariants) identifies gravitational anomalies in the field theory.

6. Outline an algorithmic approach for computing the index of an elliptic operator D using both analytic and topological methods.

Solution: The algorithm to compute the index of an elliptic operator D involves:

(a) **Input:** An elliptic operator D on a manifold M.

(b) **Step 1:** Compute the principal symbol $\sigma(D)$.

(c) **Step 2:** Verify the ellipticity by checking the invertibility of $\sigma(D)(x, \xi)$ for all non-zero ξ.

(d) **Step 3:** Determine the K-theory class $[\sigma(D)]$ in $K^0(T^*M)$ for topological index.

(e) **Step 4:** Use the heat kernel method:

$$\texttt{ind\_a}(D) = \lim_{t \to 0} \int_M \mathrm{tr}(e^{-tD^*D})(x)\,dx$$

(f) **Step 5:** Compare the analytic index with the topological data to verify or infer the topological index.

(g) **Output:** The index of D, validating both analytical and topological consistency.

Practice Problems 3

1. Explain the significance of the Atiyah-Singer Index Theorem in connecting analysis and topology. Why is it particularly important in the context of quantum field theory (QFT)? 4cm

2. Given an elliptic differential operator $D : E \to F$ on a compact manifold M, express the analytic index in terms of the kernel and cokernel of D. 4cm

3. Describe how the principal symbol $\sigma(D)$ of an elliptic operator D is used to determine ellipticity. What conditions must be satisfied? 4cm

4. Outline the role of topological K-theory in defining the topological index of an elliptic operator. How does it relate to the symbol $\sigma(D)$? 4cm

5. In the heat equation approach, how is the analytic index of an elliptic operator D evaluated using heat kernels? 4cm

6. Discuss the application of the Atiyah-Singer Index Theorem to chiral anomalies in quantum field theory. How does it assist in calculating anomalies associated with the Dirac operator? 4cm

Answers 3

1. **Solution:** The Atiyah-Singer Index Theorem is significant because it establishes a profound connection between the analytic properties of differential operators and the topological characteristics of the manifolds on which they act. Specifically, it equates the analytic index, which involves kernel and cokernel dimensions, to a topological index that is invariant under continuous deformations. In the context of QFT, this theorem is crucial as it helps understand the behavior of fermionic fields, particularly in dealing with anomalies where classical symmetries might break down upon quantization. The index theorem provides the necessary bridge to compute these anomalies, thus ensuring consistency in quantum theories.

2. **Solution:** The analytic index (`ind_a`) of an elliptic differential operator $D : E \rightarrow F$ on a compact manifold M is given by the formula:

$$\texttt{ind\_a}(D) = \dim(\ker(D)) - \dim(\mathrm{coker}(D)),$$

where $\ker(D)$ is the space of solutions to $D\phi = 0$ (kernel), and $\mathrm{coker}(D)$ is the quotient space $F/\mathrm{im}(D)$, often identified with $\ker(D^*)$, the kernel of the adjoint operator D^* (cokernel).

3. **Solution:** The principal symbol $\sigma(D)$ of an elliptic operator D plays a key role in determining ellipticity. For D to be elliptic, its principal symbol must be invertible for every point $x \in M$ and for all non-zero cotangent vectors $\xi \in T_x^*M$. Mathematically, this is expressed as:

$$\sigma(D)(x, \xi) : E_x \rightarrow F_x \text{ is invertible for all } x \in M, \xi \neq 0.$$

The invertibility of $\sigma(D)(x, \xi)$ ensures that D doesn't possess any local degeneracies and supports global solutions.

4. **Solution:** Topological K-theory is utilized to define the topological index (`ind_t`) of an elliptic operator. The principal symbol $[\sigma(D)]$ of the operator is assigned to a class in the K-theory group $K^0(T^*M)$. The topological index is derived by mapping this class to an integer via a topological index map. This process involves characteristic classes and remains invariant under homotopies, making the index a robust characterization of the operator in terms of topology.

5. **Solution:** In the heat equation approach, the analytic index of an elliptic operator D is evaluated using heat kernels derived from the associated operator D^*D. The heat operator e^{-tD^*D} is used to smooth the analysis, and the index is evaluated as:

$$\texttt{ind\_a}(D) = \lim_{t \to 0} \int_M \mathrm{tr}(e^{-tD^*D})(x)\, dx.$$

This expression considers the trace of the heat kernel and reveals the index through asymptotic limits as t approaches zero, thereby connecting with the heat content of M.

6. **Solution:** The application of the Atiyah-Singer Index Theorem to chiral anomalies in quantum field theory is instrumental. In theories involving fermions, anomalies signal a breakdown of classical symmetry at the quantum level, often represented through the Dirac operator. The index theorem helps calculate these anomalies by examining the difference between positive and negative chirality zero modes of the Dirac operator:

$$\texttt{ind\_a}(D) = n_+ - n_-,$$

where $D = \gamma^\mu(\partial_\mu + iA_\mu)$. This calculation not only quantifies the anomaly but also relates it to topological invariants, thereby ensuring that unphysical results are avoided in the quantum field theoretical setup.

Chapter 27

Anomalies in Quantum Field Theory

Introduction to Anomalies

Anomalies in quantum field theory (QFT) are phenomena where symmetries of a classical field theory are not preserved upon quantization. These anomalies play a crucial role in understanding the consistency and physical implications of quantum theories. The presence of anomalies can influence the renormalizability and the gauge invariance of a theory.

Mathematical Origin of Anomalies

Anomalies typically arise due to the failure of classical symmetry arguments when applied to quantum mechanical processes. Mathematically, they are connected to the non-invariance of certain path integrals under symmetry transformations. Consider a symmetry transformation g acting on a field ϕ in a path integral:

$$Z = \int \mathcal{D}\phi \, e^{iS[\phi]},$$

where Z is the partition function and $S[\phi]$ is the action. If Z is not invariant under g, then an anomaly is present.

1 Chiral Anomalies

Chiral anomalies are a type of anomaly associated with chiral symmetry in field theories with fermions. The divergence of the axial current J_5^μ is given by:

$$\partial_\mu J_5^\mu = \frac{e^2}{16\pi^2} \epsilon^{\mu\nu\rho\sigma} F_{\mu\nu} F_{\rho\sigma},$$

where $\epsilon^{\mu\nu\rho\sigma}$ is the Levi-Civita symbol, and $F_{\mu\nu}$ is the electromagnetic field strength tensor. The non-zero divergence on the right-hand side implies the breaking of chiral symmetry.

2 Gravitational Anomalies

Gravitational anomalies occur in theories involving fermions coupled to curved spacetime backgrounds. These anomalies manifest in the form of non-conservation of the energy-momentum tensor. Consider the energy-momentum 2-form T in a curved space:

$$d * T = \frac{c}{24\pi^2} \operatorname{tr}(R \wedge R),$$

where R is the Riemann curvature tensor, and c is a constant. Here, the right-hand side represents the anomaly.

Implications for Physical Theories

The existence of anomalies has profound implications on the theoretical landscape of QFT. Anomalies can lead to the breakdown of gauge invariance, affecting the consistency of a theory.

1 Cancellation of Anomalies

For a quantum field theory to be consistent, it is often necessary to cancel anomalies. This requirement imposes strict constraints on the possible particle content and symmetry group structure of the theory. An example of this cancellation condition is seen in the Standard Model of particle physics, where contributions to the anomaly from different fermion families cancel due to the specific structure of the model.

2 Applications to the Standard Model

In the Standard Model, anomaly cancellation is critical for maintaining gauge invariance. The cancellation ensures that the gauge fields remain massless, preserving the unbroken gauge symmetries of the model. The well-known triangle anomaly is an example where gauge bosons coupling to fermions must satisfy certain cancellation conditions:

$$\sum_i Q_i^3 = 0,$$

where Q_i are the charges of the fermions.

Algorithmic Approach to Detect Anomalies

An algorithmic approach can be employed to systematically detect potential anomalies in a given theoretical framework. The workflow involves checking the divergence of currents, evaluating path integrals under symmetry transformations, and considering charge assignments and symmetry group representations.

Algorithm 12: Anomaly Detection in Quantum Field Theories
Data: Quantum field theory with defined symmetry and fermion content
Result: Detection of anomalies
Identify all relevant symmetry currents;
Compute the divergence of each symmetry current $\partial_\mu J^\mu$;
Evaluate contribution from fermions and other fields to each current's divergence;
Summarize contributions into anomaly coefficients;
Check for cancellation conditions $\sum_i Q_i^3 = 0$ for gauge anomalies;
Confirm physical consistency if all anomalies cancel; otherwise, reevaluate the theory's setup;

Practice Problems 1

1. Explain the concept of anomaly in the context of quantum field theory. What is the significance of anomalies in terms of gauge symmetries?

2. Given a symmetry transformation g acting on a field ϕ with partition function defined by the path integral:

$$Z = \int \mathcal{D}\phi \, e^{iS[\phi]},$$

describe under what condition an anomaly arises and provide a brief explanation of why it occurs.

3. Derive the form of the divergence expression for the axial current J_5^μ in the presence of a chiral anomaly. Why is the presence of a non-zero divergence significant?

4. Discuss what is meant by anomaly cancellation in quantum field theories. Why is anomaly cancellation essential in the context of the Standard Model?

5. Identify and explain the role of the triangle anomaly in the context of gauge invariance. What mathematical condition must be satisfied for the triangle anomaly to cancel?

6. Consider a field theory with fermions and spacetime curvature. Explain how gravitational anomalies manifest and describe the equation involving the energy-momentum 2-form T.

Answers 1

1. **Solution:** Anomalies in quantum field theory represent the breakdown of a symmetry that is present at the classical level upon quantization. An anomaly signifies that the quantum version of the theory does not preserve a classical symmetry, such as gauge symmetry. This is significant because gauge symmetries are essential for the consistency and renormalizability of the theory, and the failure to maintain these symmetries can lead to unphysical results, such as violation of charge conservation.

2. **Solution:** An anomaly arises when the partition function Z, defined as

$$Z = \int \mathcal{D}\phi \, e^{iS[\phi]},$$

is not invariant under the action of a symmetry transformation g. Such non-invariance is due to the fact that the measure $\mathcal{D}\phi$ and the exponentiated action $e^{iS[\phi]}$ transform differently under g. In particular, if this transformation causes a phase shift that cannot be absorbed or ignored, it leads to the presence of an anomaly, indicating a discrepancy between classical symmetry expectations and quantum reality.

3. **Solution:** The divergence of the axial current J_5^μ is expressed as:

$$\partial_\mu J_5^\mu = \frac{e^2}{16\pi^2} \epsilon^{\mu\nu\rho\sigma} F_{\mu\nu} F_{\rho\sigma}.$$

This expression arises in the context of chiral anomalies, where the chiral symmetry is not preserved under quantum fluctuations. The non-zero divergence indicates that the classical conservation law (associated with the axial current) no longer holds at the quantum level, implying a fundamental change in the symmetry structure of the theory due to quantum effects.

4. **Solution:** Anomaly cancellation refers to the requirement that the net contribution of possible anomalies from different components of a theory must sum to zero to maintain consistency. In the Standard Model, this requirement is particularly crucial because gauge symmetries must be preserved for the theory to describe massless gauge bosons and to maintain overall consistency. Any uncancelled anomaly would result in gauge invariance being violated, potentially leading to inconsistencies and the breakdown of the predictive power of the theory.

5. **Solution:** The triangle anomaly is a specific type of anomaly that arises from the loop diagrams known as "triangle diagrams" in quantum field theories, potentially breaking gauge invariance. The mathematical condition for the triangle anomaly to cancel is:

$$\sum_i Q_i^3 = 0,$$

where Q_i are the charges of the fermions involved. This condition ensures that the contributions from different fermions to the anomaly cancel each other out, preserving gauge invariance in the model.

6. **Solution:** Gravitational anomalies in theories involving fermions in curved spacetime result in the non-conservation of the energy-momentum tensor. The corresponding equation is:

$$d * T = \frac{c}{24\pi^2} \operatorname{tr}(R \wedge R),$$

where R is the Riemann curvature tensor and c is a constant. This equation indicates that the trace portion of the energy-momentum tensor diverges due to the presence of curvature, implying that the gravitational symmetry is not perfectly preserved in quantum processing, leading to potential inconsistencies that must be resolved for a consistent quantum theory of gravity.

Practice Problems 2

1. Consider the axial current J_5^μ associated with chiral anomalies. Verify the expression for the divergence:

$$\partial_\mu J_5^\mu = \frac{e^2}{16\pi^2} \epsilon^{\mu\nu\rho\sigma} F_{\mu\nu} F_{\rho\sigma}$$

2. Show that the partition function Z is not invariant under a specific symmetry transformation g, indicating an anomaly:

$$Z = \int \mathcal{D}\phi \, e^{iS[\phi]}$$

3. Determine the anomaly present when the energy-momentum tensor in gravitational theories fails to conserve:

$$d * T = \frac{c}{24\pi^2} \operatorname{tr}(R \wedge R)$$

4. Derive the condition for the cancellation of gauge anomalies in the Standard Model:

$$\sum_i Q_i^3 = 0$$

5. Outline the algorithmic approach to detect anomalies in quantum field theories and specify the anomaly coefficients used in the detection.

6. Explain how anomalies influence the consistency of quantum field theories, particularly focusing on their impact on gauge invariance and the role of cancellation.

Answers 2

1. **Solution:** The divergence of the axial current is found using the anomaly equation:

$$\partial_\mu J_5^\mu = \frac{e^2}{16\pi^2} \epsilon^{\mu\nu\rho\sigma} F_{\mu\nu} F_{\rho\sigma}.$$

To verify this, consider the role of $\epsilon^{\mu\nu\rho\sigma}$ as the Levi-Civita symbol ensuring antisymmetry. With $F_{\mu\nu}$ as the electromagnetic field strength tensor, its presence indicates a non-zero divergence, demonstrating the breaking of chiral symmetry due to the quantum effects that cannot be canceled by classical symmetries.

2. **Solution:** A path integral formulation is not invariant under g if:

$$\delta Z = Z' - Z \neq 0.$$

For a symmetry transformation g, acting on a field ϕ, we evaluate:

$$Z' = \int \mathcal{D}\phi' \, e^{iS[\phi']} \neq \int \mathcal{D}\phi \, e^{iS[\phi]} = Z.$$

Here, the measure $\mathcal{D}\phi$ or the action $S[\phi]$ changes under g leading to an anomaly.

3. **Solution:** Gravitational anomalies arise when:

$$d * T = \frac{c}{24\pi^2} \, \text{tr}(R \wedge R).$$

Here, R is the curvature form from the Riemann curvature tensor. The form $\text{tr}(R \wedge R)$ indicates a topological term, contributing to a non-zero divergence of the energy-momentum tensor T. This signals a breakdown in energy-momentum conservation resultant from quantum gravitational effects.

4. **Solution:** Gauge anomaly cancellation requires:

$$\sum_i Q_i^3 = 0.$$

Fermions contribute to anomalies based on their charge Q_i. Each fermion's cubic charge term contributes to the anomaly coefficient. For anomaly to be canceled, these terms must sum to zero. This is crucial in models like the Standard Model, where different particle generations assure cancellation.

5. **Solution:** The algorithm to detect anomalies involves:

   (a) Identifying relevant symmetry currents.

   (b) Computing divergences $\partial_\mu J^\mu$.

   (c) Evaluating contributions from fields to divergences.

   (d) Summarizing into coefficients such as those for gauge anomalies.

   (e) Checking for cancellation conditions like $\sum_i Q_i^3 = 0$.

   (f) Confirming consistency, otherwise modifying theoretical assumptions.

6. **Solution:** Anomalies pose concerns for theory consistency:

   - They break gauge invariance which could lead to massive gauge bosons.
   - Theoretical models require anomaly cancellation to ensure unbroken symmetries at quantum level.
   - Various standard model parameters, including particle content, are tuned to achieve anomaly-free constructions.

Gauge invariance is critical for a predictive and renormalizable theory, prompting strict management of anomaly implications.

Practice Problems 3

1. Explain the significance of anomalies in quantum field theory with an emphasis on their impact on gauge invariance and renormalizability.

2. Given the path integral expression for the partition function $Z = \int \mathcal{D}\phi \, e^{iS[\phi]}$, describe mathematically how an anomaly can manifest through the non-invariance of this integral under a symmetry transformation.

3. Derive the expression for the divergence of the axial current J_5^μ highlighting the relationship with chiral anomalies.

4. Explain how gravitational anomalies can result in the non-conservation of the energy-momentum tensor using the energy-momentum 2-form T.

5. Discuss the importance of anomaly cancellation conditions in the context of the Standard Model of particle physics.

6. Outline an algorithmic approach to systematically detect anomalies in quantum field theories, explaining each step of the algorithm provided.

Answers 3

1. **Significance of Anomalies:** Anomalies are fundamental to quantum field theory as they highlight discrepancies between classical and quantum symmetries. They often signify a breakdown of gauge invariance and can impede renormalizability. For instance, gauge anomalies can lead to nonsensical predictions where gauge invariance is violated, making a theory inconsistent. Therefore, the detection and cancellation of such anomalies are critical in ensuring a viable and consistent theory.

2. **Manifestation in Path Integrals:** An anomaly manifests when a symmetry transformation g does not leave the partition function $Z = \int \mathcal{D}\phi\, e^{iS[\phi]}$ invariant. For instance, if under g, the measure changes by a phase factor that cannot be reabsorbed, such as in the Jacobian determinant, the integral is not invariant, indicating a quantum anomaly.

3. **Divergence of Axial Current:** For chiral anomalies, the divergence of the axial current J_5^μ is influenced by quantum effects, effective in gauge theories with fermions. The relation:

$$\partial_\mu J_5^\mu = \frac{e^2}{16\pi^2} \epsilon^{\mu\nu\rho\sigma} F_{\mu\nu} F_{\rho\sigma},$$

indicates that quantum corrections lead to a non-zero divergence due to topological effects in gauge field backgrounds, breaking chiral symmetry.

4. **Non-Conservation of Energy-Momentum Tensor:** In curved spacetimes, gravitational anomalies result in an anomalous term in the conservation law of the energy-momentum tensor. Consider:

$$d * T = \frac{c}{24\pi^2} \operatorname{tr}(R \wedge R),$$

illustrating that quantum effects and topological terms in curvature (via trace) prevent T from being divergence-free, representing the gravitational anomaly.

5. **Importance in Standard Model:** Anomalies, particularly gauge anomalies, must cancel to maintain gauge invariance in the Standard Model. This cancellation is achieved by balancing fermion contributions to the anomalies, ensuring gauge consistency and masslessness of gauge fields. For instance, anomaly cancellation conditions ensure the strength and structure of interactions in the Standard Model.

6. **Algorithmic Approach to Anomaly Detection:** The steps involved are:

   - Identify symmetry currents;
   - Compute divergences of these currents;
   - Calculate contributions from fermions to each current's divergence;
   - Sum contributions as coefficients;
   - Check for specific conditions (e.g., third power of the fermion charge sum to zero for gauge anomalies);
   - Confirm consistency by ensuring no net anomaly or revise the theoretical construction.

This algorithmic approach efficiently tests for anomalies, ensuring a consistent quantum framework.

Chapter 28

Non-Perturbative Methods in Functional Analysis

Introduction to Non-Perturbative Techniques

Non-perturbative methods involve calculating physical quantities without relying on the perturbation theory's expansion in small parameters. These techniques are crucial in understanding phenomena where perturbative approaches fail, such as strong coupling scenarios.

Operator Algebras in Quantum Field Theory

Operator algebras provide a robust mathematical framework for non-perturbative analysis. Two key types are considered: C*-algebras and von Neumann algebras.

1 C*-Algebras

C*-algebras are algebraic structures composed of operators on a Hilbert space. They are equipped with operations such as addition, multiplication, and an involution operation that resembles taking adjoints. The defining property is:

$$\|a^*a\| = \|a\|^2$$

for any a in the algebra, where a^* denotes the adjoint of a.

2 Von Neumann Algebras

Von Neumann algebras are a subset of C*-algebras, characterized by their closure in the weak operator topology. They are often utilized to model quantum mechanical observables in quantum field theory. For a von Neumann algebra $\mathcal{M}$, it holds that:

$$\mathcal{M} = \mathcal{M}''$$

where $\mathcal{M}''$ is the double commutant of $\mathcal{M}$.

Topological Methods in Quantum Field Theory

Topological methods provide tools to study the global properties of spaces and fields, crucial for non-perturbative analysis.

1 Fiber Bundles

Fiber bundles offer a topological structure where a space E (the total space) is mapped onto a base space B via a projection $\pi : E \to B$. Locally, E looks like a product of B and a typical fiber F. This is mathematically expressed as:

$$\pi^{-1}(U) \cong U \times F$$

for an open set $U \subset B$.

2 Homotopy and Cohomology

Homotopy groups and cohomology rings serve as topological invariants. These concepts describe the ability to deform one map into another and the structure of differential forms over a space. For a map $f : X \to Y$, two maps are homotopic if there exists a continuous transformation H such that:

$$H(x,0) = f(x), \quad H(x,1) = g(x) \quad \forall x \in X$$

The cohomology ring $H^*(X)$ aggregates cohomology groups, with cup products providing an algebraic structure.

Applications in Small-Coupling and Strong-Coupling Regimes

Non-perturbative techniques are employed in both small-coupling and strong-coupling regimes to address limitations of perturbative methods.

1 Small-Coupling Regimes

In small-coupling scenarios, while perturbative expansions are generally valid, non-perturbative corrections can be significant. An example involves instantons, which are solutions to classical field equations contributing to quantum processes. Mathematically, a general instanton contribution ΔS can be expressed as:

$$\Delta S = \exp\left(-\frac{1}{g^2} S_{\text{inst}}\right)$$

where S_{inst} is the instanton action, and g is the coupling constant.

2 Strong-Coupling Regimes

Strong-coupling regimes inherently defy perturbative expansions. Non-perturbative methods like lattice gauge theory become essential. Lattice gauge theory discretizes spacetime into a lattice, enabling numerical simulations. The partition function on a lattice can be expressed as:

$$Z = \int \prod_{x,\mu} dU_\mu(x)\, e^{-S[U]}$$

where $U_\mu(x)$ are link variables, and $S[U]$ is the action in terms of these variables.

Algorithm for Non-Perturbative Analysis

Algorithm 13: Non-Perturbative Analysis Strategy
Data: Quantum field theory model
Result: Evaluate non-perturbative effects
Identify symmetries and construct operator algebra;
Classify topological features using homotopy and cohomology;
Calculate physical quantities using path integral methods ignoring small parameter expansions;
Use lattice methods for strong-coupling calculations;
Validate results through symmetry and topological consistency checks;

Practice Problems 1

1. Define a C*-algebra and show that for any element a in a C*-algebra, it holds that:

$$\|a^*a\| = \|a\|^2$$

2. What is the significance of the equality $\mathcal{M} = \mathcal{M}''$ in von Neumann algebras?

3. Describe the structure of a fiber bundle and explain the meaning of the local triviality condition:

$$\pi^{-1}(U) \cong U \times F$$

4. Explain how instantons can contribute to quantum processes in small-coupling regimes with a specific example using the instanton action:

$$\Delta S = \exp\left(-\frac{1}{g^2} S_{\text{inst}}\right)$$

5. Discuss the role of lattice gauge theory in studying strong-coupling regimes and how the partition function is represented:

$$Z = \int \prod_{x,\mu} dU_\mu(x)\, e^{-S[U]}$$

6. Outline the key steps of the non-perturbative analysis strategy algorithm presented in the chapter and discuss its significance.

Answers 1

1. **Defining C*-Algebra:** A C*-algebra is a complex algebra $\mathcal{A}$ of bounded linear operators on a Hilbert space $\mathcal{H}$ with:
   - an involution $* : \mathcal{A} \to \mathcal{A}$, for $a \in \mathcal{A}$, a^* is the adjoint of a.
   - a norm $\|\cdot\|$ satisfying $\|a^* a\| = \|a\|^2$.

   Proof: The C*-algebra norm satisfies $\|a^* a\| = \|a\|^2$ by definition. Recall:

   $$\|a\|^2 = \sup_{\|x\|=1} \langle a^* a x, x \rangle = \|a^* a\|$$

Therefore, the norm condition $\|a^*a\| = \|a\|^2$ holds, confirming a's properties in this algebraic structure.

2. **Significance of $\mathcal{M} = \mathcal{M}''$:** This equality characterizes von Neumann algebras, signifying that the algebra is closed in the weak operator topology and equals its double commutant, representing all the bounded operators preserving quantum mechanical observables' completion.

3. **Structure of Fiber Bundles:** A fiber bundle consists of:
   - Total space E
   - Base space B
   - Typical fiber F
   - Projection $\pi : E \to B$

   Local Triviality Condition: Locally over an open set $U \subset B$:

   $$\pi^{-1}(U) \cong U \times F$$

   This condition implies that each point in U is associated with a fiber, similar to a local product space, enabling complex structures' study and global properties analysis.

4. **Instanton Contribution Example:** Instantons contribute to quantum processes:
   - Non-perturbative corrections are significant despite small coupling.
   - Example: In Yang-Mills theory, instantons represent classical solutions affecting processes.

   Calculating Contribution: Using instanton action,

   $$\Delta S = \exp\left(-\frac{1}{g^2} S_{\text{inst}}\right)$$

   This demonstrates the exponentially suppressed contributions, evident in tunneling phenomena and vacua structure in quantum calculations.

5. **Role of Lattice Gauge Theory:** In strong-coupling regimes, lattice gauge theory:
   - Discretizes spacetime for numerical studies.
   - Helps approximate non-perturbative effects on a lattice grid.

   Partition Function Representation:

   $$Z = \int \prod_{x,\mu} \mathrm{d}U_\mu(x) \, e^{-S[U]}$$

   Describes the sum over link variables $U_\mu(x)$, key to understanding gauge fields' configuration space at strong coupling through simulations.

6. **Non-Perturbative Analysis Strategy:**

   (a) Identify symmetries and construct operator algebra.

   (b) Classify topological features using homotopy and cohomology.

   (c) Evaluate physical quantities avoiding small parameter expansions.

   (d) Use lattice methods for strong-coupling scenarios.

   (e) Validate through symmetry and topological consistency checks.

   Significance: This strategy is crucial for exploring phenomena where perturbation theory fails, offering robust methods for quantitative predictions, crucial to understanding the underlying physics of quantum field theories.

Practice Problems 2

1. Explain why non-perturbative methods are necessary for analyzing strong-coupling regimes in field theory.

2. Show that the defining property of a C*-algebra, $\|a^*a\| = \|a\|^2$, implies that the norm is submultiplicative, i.e., $\|ab\| \leq \|a\|\|b\|$ for any elements a, b in the algebra.

3. Prove that von Neumann algebras are closed under the weak operator topology.

4. Describe how fiber bundles can be used to model topological features in quantum field theory, providing an example related to gauge theories.

5. Outline the role of lattice gauge theory in non-perturbative analysis and derive the expression for the partition function $Z = \int \prod_{x,\mu} dU_\mu(x) \, e^{-S[U]}$ on a lattice.

6. Discuss the mathematical significance of homotopy groups in the context of quantum field theory.

Answers 2

1. **Solution:** Non-perturbative methods are crucial in strong-coupling regimes because in such cases, the coupling constant is not small, and perturbative expansions based on small parameters become unreliable. For example, in Quantum Chromodynamics (QCD), the strong force coupling becomes significant at low energies, necessitating approaches that can handle such strong interactions without relying on expansions in small parameters. Non-perturbative methods, such as lattice gauge theory, allow for numerical solutions that describe these scenarios accurately.

2. **Solution:** The defining property $\|a^*a\| = \|a\|^2$ implies the submultiplicative property of the norm:

$$\|ab\|^2 = \|(ab)^*(ab)\| = \|b^*a^*ab\| \leq \|b^*\|\|a^*a\|\|b\|$$

Since $\|a^*a\| = \|a\|^2$, we obtain:
$$\|ab\|^2 \leq \|b\|^2\|a\|^2$$

Taking the square root:
$$\|ab\| \leq \|a\|\|b\|$$

Hence, the norm is submultiplicative.

3. **Solution:** By definition, von Neumann algebras are closed in the weak operator topology, which means they contain all limit points of sequences of operators converging weakly. Consider a sequence $\{T_n\}$ in von Neumann algebra $\mathcal{M}$ such that $T_n \to T$ weakly. For each vector x and all vectors y, $\langle T_n x, y \rangle \to \langle Tx, y \rangle$. Since collections of polynomials in operators are dense, and weak closures of algebras are larger only if closure is added, T is within $\mathcal{M}$, confirming that it is weakly closed.

4. **Solution:** Fiber bundles describe fields in quantum field theory, particularly gauge fields. In a gauge theory, such as electromagnetism, a principal bundle with a structure group (e.g., $U(1)$ for electromagnetism) models the gauge symmetry. Locally, the bundles represent the relation between space (base space B) and field values (fiber F). For instance, the electromagnetic potential can be seen as a connection on a $U(1)$ bundle, allowing transformation globally consistent with local gauge invariance.

5. **Solution:** Lattice gauge theory discretizes spacetime to approximate path integrals for field theories. The partition function represents the statistical sum over all observable states, expressed as:

$$Z = \int \prod_{x,\mu} dU_\mu(x) \, e^{-S[U]}$$

where $U_\mu(x)$ are link variables (group elements of the gauge group associated with lattice links); $S[U]$ is the lattice action, a discretized version of the field theory action.

6. **Solution:** Homotopy groups classify topological spaces based on paths. In QFT, they are used to determine whether different configurations can be continuously deformed into each other. For example, the first homotopy group (π_1) tells us about the existence of topologically distinct monopoles or solitons, which can't be obtained through simple perturbations and represent non-perturbative features.

Practice Problems 3

1. Explain the significance of C*-algebras in non-perturbative approaches to quantum field theory, particularly in terms of their structure and properties.

2. Demonstrate how von Neumann algebras are utilized to model quantum mechanical observables, particularly focusing on the importance of the weak operator topology.

3. Describe the role of fiber bundles in the topological study of quantum field theories. Provide an example of how a fiber bundle can be applied to gauge theories.

4. How do instantons provide insights into non-perturbative phenomena in quantum field theory? Include a mathematical expression involving an instanton contribution and explain its components.

5. Explain the process and significance of applying lattice gauge theory in strong-coupling regimes. How is the partition function formulated in this method?

6. Describe an algorithmic approach to evaluating non-perturbative effects in quantum field theory and provide a step-by-step explanation of each step involved.

Answers 3

1. **Solution:** C*-algebras are vital for non-perturbative approaches because they offer a framework for dealing with operators on Hilbert spaces. The operations they include (addition, multiplication, and involution) maintain the algebra's structure under limits, making them suitable for studying continuous symmetries and observables in quantum mechanics and quantum field theory. The defining property $\|a^*a\| = \|a\|^2$ ensures this structure remains normed and facilitates the spectral analysis of operators.

2. **Solution:** Von Neumann algebras model quantum mechanical observables by being closed in the weak operator topology, which addresses the convergence problems typical in quantum theories. These algebras' property $\mathcal{M} = \mathcal{M}''$ emphasizes their rich algebraic structure, accommodating infinite-dimensional spaces encountered in quantum field theory. The weak topology facilitates studying these algebras' dual spaces and spectral properties, crucial for analyzing quantum observables.

3. **Solution:** Fiber bundles are foundational in topological studies as they allow the modeling of complex multi-parameter spaces as products of simpler spaces, aiding in understanding fields over a base space. In gauge theories, a fiber bundle can describe how local gauge transformations relate to global field configurations. The expression $\pi^{-1}(U) \cong U \times F$ shows how fibers (local symmetries) bundle over a base space, illustrating the concept of gauge transformations affecting fields.

4. **Solution:** Instantons shed light on phenomena like tunneling effects in non-perturbative quantum field theory by contributing to path integrals with significant corrections. The expression $\Delta S = \exp\left(-\frac{1}{g^2} S_{\text{inst}}\right)$ indicates an exponential term heavily impacting the system at small coupling constants g. Here, S_{inst} is the classical solution's action, which directly influences quantum corrections, revealing deep insights into transitions and symmetry breaking.

5. **Solution:** Lattice gauge theory, applied in strong-coupling scenarios, discretizes spacetime into a lattice to permit numerical simulations. This approach circumvents the non-convergence of perturbation series by evaluating path integrals directly within a finite, tractable system. The partition function, $Z = \int \prod_{x,\mu} dU_\mu(x)\, e^{-S[U]}$, where $U_\mu(x)$ represents link variables and $S[U]$ dictates the lattice's action, illustrates this methodological shift from continuum to discrete computational frameworks.

6. **Solution:** An algorithm for non-perturbative analysis involves several steps:

   (a) Identify symmetries and construct associated operator algebras to understand the algebraic structure of the theory.
   (b) Classify topological features via homotopy and cohomology to address global properties.
   (c) Calculate physical quantities using path integrals, avoiding perturbative series.
   (d) Use lattice methods for strong-coupling calculations, ensuring computational feasibility.
   (e) Validate results through symmetry and topological consistency checks, ensuring that the findings align with theoretical predictions.

   This structured approach ensures comprehensive exploration of non-perturbative phenomena.

Chapter 29

Thermal Field Theory and KMS Condition

Introduction to Thermal Quantum Field Theory

In the framework of thermal quantum field theory (QFT), the interactions of quantum fields at finite temperature are studied. This discipline extends zero-temperature quantum field theory by incorporating statistical mechanics principles, which is essential for understanding systems in thermal equilibrium.

Statistical Mechanics in Quantum Field Theory

Statistical mechanics in the context of QFT is a natural extension where the goal is to describe the macroscopic properties of a system based on its microscopic components. At thermal equilibrium, a quantum field theory is often described by the partition function Z, given by

$$Z = \text{Tr}\left(e^{-\beta H}\right)$$

where $\beta = \frac{1}{k_B T}$, with k_B as the Boltzmann constant, T as temperature, and H as the Hamiltonian operator of the system. The partition function serves as a generating functional from which various thermodynamic quantities can be derived.

The KMS Condition

The Kubo-Martin-Schwinger (KMS) condition characterizes thermal equilibrium states in quantum statistical mechanics. For a given observance A and B, the equilibrium condition is expressed in terms of correlation functions.

Given a state ω and two operators A and B, the KMS condition is:

$$\omega(A(t)B) = \omega(BA(t + i\beta))$$

for any real t, where $A(t) = e^{iHt} A e^{-iHt}$. This reflection property over imaginary time $i\beta$ is a cornerstone for identifying thermal equilibrium states.

1 Implications of the KMS Condition

The KMS condition implies that for any two observables A and B, the expression

$$\langle A(t)B \rangle_\beta = \langle BA(t + i\beta) \rangle_\beta$$

holds, establishing a relation between time correlation functions at different times and ensuring consistency with equilibrium statistical mechanics.

Thermal Quantum Field Theory Framework

Within thermal field theory, finite-temperature effects are incorporated by performing a Wick rotation in time to imaginary values, which transforms the Minkowski space formalism into Euclidean space. This analytic continuation into imaginary time leads to a compactified time dimension of length β, resulting in a rich interplay between field theory and statistical ensembles.

1 Path Integral Formulation

In the path integral formalism, the partition function Z is represented as a functional integral over all field configurations:

$$Z = \int \mathcal{D}[\phi]\, e^{-S_E[\phi]}$$

Here, $S_E[\phi]$ denotes the Euclidean action, obtained by the analytic continuation of the classical action from Minkowski space to Euclidean space via $t \to -i\tau$.

2 Thermal Propagators

Thermal propagators in quantum field theory provide insights into particle interactions at finite temperature. The free thermal propagator for a bosonic field, confined to imaginary time, is given by:

$$\Delta(\tau, \mathbf{x}) = \frac{1}{\beta} \sum_{n=-\infty}^{+\infty} \int \frac{d^3\mathbf{p}}{(2\pi)^3} \frac{e^{i(\mathbf{p}\cdot\mathbf{x} - \omega_n \tau)}}{\omega_n^2 + \mathbf{p}^2 + m^2}$$

where $\omega_n = \frac{2\pi n}{\beta}$ are the Matsubara frequencies specific to bosons.

3 Fermionic Systems

For fermionic fields, the Matsubara frequencies differ by a half-integer, accounting for the antiperiodicity condition in imaginary time:

$$\omega_n = \frac{(2n+1)\pi}{\beta}$$

The thermal propagator is similar in form to the bosonic case, albeit with these adjusted frequencies and accounting for the particle's spin-statistics.

Algorithm for Calculating Thermal Green's Functions

The evaluation of thermal Green's functions involves the implementation of path integrals in a thermal setting.

Algorithm 14: Thermal Green's Functions Calculation

Data: Quantum field theory model at finite temperature
Result: Compute thermal Green's functions
Perform Wick rotation: Replace t with $-i\tau$;
Define the path integrals over Euclidean actions;
Implement Matsubara frequency sums reflecting periodicity/antiperiodicity;
Evaluate the resulting integrals over spatial components;
Obtain expressions for thermal Green's functions to analyze finite-temperature effects;

Practice Problems 1

1. Explain how the partition function $Z = \text{Tr}\left(e^{-\beta H}\right)$ is related to observable macroscopic quantities in a quantum field theory at thermal equilibrium.

2. Derive the Kubo-Martin-Schwinger (KMS) condition starting from the definition of thermal states in statistical mechanics.

3. Show how the Matsubara frequencies $\omega_n = \frac{2\pi n}{\beta}$ in bosonic fields arise from performing the Wick rotation.

4. Describe the significance of the Wick rotation in transforming a quantum field theory from Minkowski space to Euclidean space.

5. Calculate the thermal propagator for a fermionic field, given the antiperiodicity condition in imaginary time.

6. Explain how the path integral formulation helps in computing thermal Green's functions and what role the Euclidean action plays in this framework.

Answers 1

1. **Solution:**
   The partition function $Z = \text{Tr}\left(e^{-\beta H}\right)$ acts as a generating functional for macroscopic thermodynamic quantities in thermal QFT. Specifically, the free energy F is related by $F = -\frac{1}{\beta}\ln Z$. From F, other quantities like the internal energy U, entropy S, and specific heat can be derived using thermodynamic identities, such as $U = -\frac{\partial \ln Z}{\partial \beta}$, and $S = \beta(U - F)$.

2. **Solution:**
   The KMS condition is derived from the periodic boundary conditions imposed by thermal equilibrium in the path integral approach. For a thermal state characterized by $\omega(A(t)B)$ and B, using the cyclic property of the trace one has:

   $$\omega(A(t)B) = \frac{\text{Tr}\left(e^{-\beta H}e^{iHt}Ae^{-iHt}B\right)}{Z}$$

   Using $e^{-\beta H} = e^{-iH(i\beta)}$, invoking the analyticity in imaginary time $i\beta$ leads to:

   $$\omega(A(t)B) = \omega(BA(t + i\beta))$$

3. **Solution:**
   Performing a Wick rotation transforms real time t into imaginary time $\tau = it$ which is periodic with β. This requires fields to satisfy periodic boundary conditions, leading to the discrete Matsubara frequencies for bosons:

   $$\omega_n = \frac{2\pi n}{\beta}$$

   This arises because fields are periodic with period β, resulting in Fourier modes characterized by these frequencies.

4. **Solution:**
   Wick rotation $(t \to -i\tau)$ changes the theory from Minkowski to Euclidean space, converting the path integral into a statistical (thermal) partition function. This transformation facilitates easier mathematical treatment as Euclidean actions reduce to problems in statistical physics where energy minimization replaces path extremization, creating a connection between quantum mechanics and statistical mechanics.

5. **Solution:**
   For fermions, the antiperiodic condition means the field changes sign upon β shifts: $\psi(\tau + \beta) = -\psi(\tau)$. Hence, the fermionic Matsubara frequencies are:

   $$\omega_n = \frac{(2n + 1)\pi}{\beta}$$

The propagator thus involves these adjusted frequencies, represented by:

$$S(\tau, \mathbf{x}) = \frac{1}{\beta} \sum_n \int \frac{d^3p}{(2\pi)^3} \frac{e^{i(\mathbf{p} \cdot \mathbf{x} - \omega_n \tau)}}{i\omega_n - \mathbf{p} \cdot \mathbf{p} - m^2}$$

6. **Solution:**

Path integrals capture statistical properties of fields over all configurations by integrating with respect to the exponential of the Euclidean action $S_E[\phi]$. The thermal Green's functions are then average field configurations weighted by this action. The Euclidean action facilitates this by embodying both kinetic and interaction terms, encompassing the dynamics within the path integral framework and leading to calculable expressions for field correlation functions.

Practice Problems 2

1. Derive the expression for the partition function in the framework of thermal quantum field theory, expressing it in terms of the statistical mechanics principles.

   Hint: Consider the trace over the exponential of the negative inverse temperature times the Hamiltonian.

2. Explain the KMS condition for a state ω and two operators A and B. Derive its implication for the correlation functions.

   Hint: Use the time evolution of operators to express the KMS condition.

3. Describe how thermal propagators are affected by the incorporation of Matsubara frequencies in the bosonic field. Give the complete expression for the free thermal propagator.

   Hint: Consider the imaginary time formalism and sum over discrete frequencies.

4. Dissect the process of applying Wick rotation in thermal QFT and its impact on the transition from Minkowski to Euclidean space.

Hint: Analyze the transformation from real to imaginary time

and its implications on the path integral formulation.

5. Contrast the treatment of fermionic and bosonic systems at finite temperature, highlighting the differences in Matsubara frequencies.

Hint: Pay special attention to the boundary conditions in imaginary time for each type of particle.

6. Discuss how the path integral formulation is used to evaluate thermal Green's functions and outline the steps involved in this calculation.

Hint: Focus on the integral over field configurations and the role of Euclidean actions.

Answers 2

1. Derive the expression for the partition function in the framework of thermal quantum field theory.
   Solution: The partition function Z in thermal quantum field theory is given by the trace:

$$Z = \text{Tr}\left(e^{-\beta H}\right)$$

Step 1: Recognize that $\beta = \frac{1}{k_B T}$ where k_B is the Boltzmann constant and T is the temperature.

Step 2: The Hamiltonian H represents the total energy of the system.

Step 3: This exponential operator accounts for the statistical weighting of states, and taking trace essentially sums over all possible states of the system.

Step 4: This partition function acts as the generating functional, from which thermodynamic properties can be derived, such as internal energy and entropy. Therefore, the expression for the partition function is:

$$Z = \mathrm{Tr}\left(e^{-\beta H}\right)$$

2. Explain the KMS condition for a state ω and two operators A and B. **Solution:** The KMS condition is:

$$\omega(A(t)B) = \omega(BA(t + i\beta))$$

Step 1: Define $A(t) = e^{iHt} A e^{-iHt}$, using time-evolution under the Hamiltonian H.

Step 2: Recognize $A(t + i\beta)$ involves evolution through imaginary time, implying a temperature dependence.

Step 3: Rearrange the KMS condition to acquire correlation functions:

$$\langle A(t)B \rangle_\beta = \langle BA(t + i\beta) \rangle_\beta$$

Step 4: Interpreting this, the KMS condition reflects equilibrium states by flipping $i\beta$ and ensuring consistency over β as the thermal period.

3. Describe how thermal propagators are affected by the incorporation of Matsubara frequencies. **Solution:** For bosonic fields, the propagator is:

$$\Delta(\tau, \mathbf{x}) = \frac{1}{\beta} \sum_{n=-\infty}^{+\infty} \int \frac{d^3\mathbf{p}}{(2\pi)^3} \frac{e^{i(\mathbf{p}\cdot\mathbf{x} - \omega_n \tau)}}{\omega_n^2 + \mathbf{p}^2 + m^2}$$

Step 1: Define Matsubara frequencies for bosons as $\omega_n = \frac{2\pi n}{\beta}$.

Step 2: The thermal propagator sums over discrete values ω_n, reflecting periodicity.

Step 3: Analyze free bosonic field: represent imaginary time propagation via these frequencies.

Step 4: Integrate over momentum space $\mathbf{p}$ for spatial components of field propagation.

4. Dissect the process of applying Wick rotation in thermal QFT. **Solution:** Wick rotation involves transforming time to imaginary values:

$$t \to -i\tau$$

Step 1: Rotate real time t to imaginary time τ, enabling Euclidean formalism in path integrals.

Step 2: This transformation results in metric change from Minkowski (

Step 3: Euclidean space allows compactification: time dimension's length becomes β.

Step 4: Redefine path integral formulation, with integration over Euclidean actions S_E.

5. Contrast the treatment of fermionic and bosonic systems at finite temperature. **Solution:** Bosonic frequencies: $\omega_n = \frac{2\pi n}{\beta}$, periodic.

Fermionic frequencies: $\omega_n = \frac{(2n+1)\pi}{\beta}$, anti-periodic.

Step 1: Enumerate bosonic frequencies from even integer n.

Step 2: Enumerate fermionic frequencies from odd half-integers, ensuring anti-periodicity using the Fermi-Dirac statistics.

Step 3: Analyze impact on thermal propagators: periodicity differs, bounding field behavior in imaginary time differently.

6. Discuss how the path integral formulation is used to evaluate thermal Green's functions. **Solution:** The path integral for partition function:

$$Z = \int \mathcal{D}[\phi]\, e^{-S_E[\phi]}$$

Step 1: Implement Wick rotation, enabling Euclidean action $S_E[\phi]$.

Step 2: Define the thermal path integral over field configurations $[phi]$.

Step 3: Include Matsubara frequency sums: results in discrete imaginary-time boundary conditions.

Step 4: Evaluate Thermodynamic propagators through these integrals to understand finite-temperature interactions.

Practice Problems 3

1. Derive the partition function for a quantum field theory in thermal equilibrium and express it in terms of the Hamiltonian operator H:

$$Z = \mathrm{Tr}\left(e^{-\beta H}\right)$$

2. Demonstrate the Wick rotation transformation in the thermal QFT framework, from Minkowski to Euclidean space:

$$t \to -i\tau$$

3. Show that the Matsubara frequencies for bosons and fermions are given by:

$$\omega_n = \frac{2\pi n}{\beta} \quad \text{(bosons)}, \quad \omega_n = \frac{(2n+1)\pi}{\beta} \quad \text{(fermions)}$$

4. Solve for the thermal propagator of a bosonic field in imaginary time:

$$\Delta(\tau, \mathbf{x}) = \frac{1}{\beta} \sum_{n=-\infty}^{+\infty} \int \frac{d^3\mathbf{p}}{(2\pi)^3} \frac{e^{i(\mathbf{p}\cdot\mathbf{x} - \omega_n \tau)}}{\omega_n^2 + \mathbf{p}^2 + m^2}$$

5. Discuss the significance of the KMS condition:

$$\omega(A(t)B) = \omega(BA(t + i\beta))$$

6. Explain the analytic continuation to Euclidean space and its impact on constructing the partition function in QFT:

$$Z = \int \mathcal{D}[\phi] \, e^{-S_E[\phi]}$$

Answers 3

1. Derive the partition function for a quantum field theory in thermal equilibrium:

   Solution: The partition function Z for a system in thermal equilibrium is calculated using the Hamiltonian operator H, defined as:

   $$Z = \text{Tr}\left(e^{-\beta H}\right)$$

   where $\beta = \frac{1}{k_B T}$, with k_B being the Boltzmann constant and T the temperature. The trace operation sums over the exponential factors of energy states, effectively describing the statistical distribution of these states under thermal conditions.

2. Demonstrate the Wick rotation transformation in thermal QFT:

   Solution: Wick rotation is a key step in transforming a theory from Minkowski to Euclidean space. This involves replacing the time variable t with an imaginary time τ, defined as:

   $$t \to -i\tau$$

   This substitution converts the metric from a Lorentzian to a Euclidean form, simplifying the treatment of finite temperature QFT as a statistical mechanics problem.

3. Show the Matsubara frequencies for bosons and fermions:

   Solution: The Matsubara frequencies arise when performing Fourier transforms in periodic (or antiperiodic) imaginary time conditions.
   - For bosons, which obey periodic boundary conditions in imaginary time:

   $$\omega_n = \frac{2\pi n}{\beta}, \quad n \in \mathbb{Z}$$

   - For fermions, due to antiperiodic boundary conditions:

   $$\omega_n = \frac{(2n+1)\pi}{\beta}, \quad n \in \mathbb{Z}$$

4. Solve for the thermal propagator of a bosonic field:

   Solution: The thermal propagator $\Delta(\tau, \mathbf{x})$ is derived by summing over all Matsubara frequencies and integrating over spatial momenta:

   $$\Delta(\tau, \mathbf{x}) = \frac{1}{\beta} \sum_{n=-\infty}^{+\infty} \int \frac{d^3\mathbf{p}}{(2\pi)^3} \frac{e^{i(\mathbf{p}\cdot\mathbf{x} - \omega_n \tau)}}{\omega_n^2 + \mathbf{p}^2 + m^2}$$

   This result utilizes the periodicity in imaginary time given by the Matsubara frequencies for bosons.

5. Discuss the significance of the KMS condition:

   Solution: The KMS condition is pivotal in identifying thermal equilibrium states. Defined through correlations as:
   $$\omega(A(t)B) = \omega(BA(t + i\beta))$$

   This condition ensures that the time evolution of an observable within a thermal state behaves consistently across imaginary time shifts of $i\beta$, reflecting periodicity in thermal equilibrium.

6. Explain the analytic continuation to Euclidean space and its impact on constructing the partition function:

   Solution: Analytic continuation involves the transformation from real to imaginary time creating a compactified time dimension. This affects the partition function, expressed in the path integral formalism:
   $$Z = \int \mathcal{D}[\phi]\, e^{-S_E[\phi]}$$

   where $S_E[\phi]$ is the Euclidean action. This modification allows for a mathematical treatment akin to statistical mechanics, making it easier to handle thermal quantum field theories.

Chapter 30

Quantum Fields on Curved Spacetimes

Introduction to Quantum Fields and Curved Spacetimes

The study of quantum fields on curved spacetimes extends the framework of quantum field theory (QFT) by incorporating the effects of spacetime curvature, as described by general relativity. This integration is essential for understanding how quantum fields behave in the presence of a gravitational field. The notion of a flat Minkowski spacetime is generalized to a curved manifold, influencing field equations and quantization processes.

Mathematical Foundations

Consider a spacetime represented by a smooth four-dimensional manifold M equipped with a metric tensor $g_{\mu\nu}$. The action of a scalar field ϕ in this curved spacetime is given by

$$S = \int d^4x \sqrt{-g} \left(\frac{1}{2} g^{\mu\nu} \nabla_\mu \phi \nabla_\nu \phi - \frac{1}{2} m^2 \phi^2 \right), \tag{30.1}$$

where ∇_μ denotes the covariant derivative compatible with $g_{\mu\nu}$, g is the determinant of the metric tensor, and m represents the mass of the scalar field.

1 The Covariant Derivative

The covariant derivative is a generalization of the partial derivative to curved spacetimes, maintaining tensorial nature under coordinate transformations. For a scalar field ϕ, the covariant derivative reduces to the ordinary derivative:

$$\nabla_\mu \phi = \partial_\mu \phi. \tag{30.2}$$

However, for a vector field V^ν, it involves the Christoffel symbols $\Gamma^\lambda_{\mu\nu}$ defined by

$$\Gamma^\lambda_{\mu\nu} = \frac{1}{2} g^{\lambda\sigma} \left(\partial_\mu g_{\sigma\nu} + \partial_\nu g_{\sigma\mu} - \partial_\sigma g_{\mu\nu} \right), \tag{30.3}$$

such that

$$\nabla_\mu V^\nu = \partial_\mu V^\nu + \Gamma^\nu_{\mu\sigma} V^\sigma. \tag{30.4}$$

Field Quantization in Curved Spacetimes

The quantization of fields in curved spacetimes poses distinct challenges due to the lack of a global time-like Killing vector field which complicates the definition of particles. Nevertheless, the canonical quantization approach can be adapted to curved spacetime.

1 Mode Decomposition

Scalar fields $\phi(x)$ are decomposed into a basis of mode functions appropriate for the geometry of the spacetime:

$$\phi(x) = \sum_i \left(a_i u_i(x) + a_i^\dagger u_i^*(x) \right), \tag{30.5}$$

where a_i and $a_i^\dagger$ are the annihilation and creation operators, respectively, satisfying the usual commutation relations:

$$[a_i, a_j^\dagger] = \delta_{ij}, \quad [a_i, a_j] = [a_i^\dagger, a_j^\dagger] = 0. \tag{30.6}$$

2 Curved Spacetime Effects

In curved spacetimes, the choice of vacuum state is nontrivial due to the lack of preferred timelike symmetry. The particle concept becomes coordinate-dependent, leading to phenomena such as Hawking radiation and the Unruh effect.

Examples of Quantum Fields in Curved Spacetimes

1 Schwarzschild Spacetime

In the Schwarzschild metric, representing the spacetime outside a spherical non-rotating mass, the metric takes the form:

$$ds^2 = -\left(1 - \frac{2M}{r}\right) dt^2 + \left(1 - \frac{2M}{r}\right)^{-1} dr^2 + r^2 d\Omega^2, \tag{30.7}$$

where $d\Omega^2$ represents the angular part of the metric. The field equations and mode expansion are influenced by this curvature, crucial in studying black hole thermodynamics.

2 Friedmann-Lemaître-Robertson-Walker (FLRW) Universe

For cosmological applications, the FLRW metric describes an expanding universe:

$$ds^2 = -dt^2 + a(t)^2 \left(\frac{dr^2}{1 - kr^2} + r^2 d\Omega^2 \right), \tag{30.8}$$

with $a(t)$ as the scale factor and k as the curvature constant. Quantum fields in this background explore the evolution of early universe phenomena, such as inflation and primordial perturbations.

The extension of quantum fields to curved spacetimes necessitates a nuanced understanding of relativistic effects and geometry, providing insights into fields interacting with gravitational fields.

Practice Problems 1

1. Show that the action of a scalar field in curved spacetime, given by

$$S = \int d^4x \sqrt{-g} \left(\frac{1}{2} g^{\mu\nu} \nabla_\mu \phi \nabla_\nu \phi - \frac{1}{2} m^2 \phi^2 \right),$$

   is invariant under coordinate transformations.

2. Derive the expression for the covariant derivative of a vector field V^ν in curved spacetime using the definition of Christoffel symbols.

3. Determine the commutation relation for scalar field operators a_i and $a_i^\dagger$ in curved spacetime given in mode decomposition:

$$\phi(x) = \sum_i \left(a_i u_i(x) + a_i^\dagger u_i^*(x) \right).$$

4. Explain how the concept of vacuum state in curved spacetime differs from that in Minkowski spacetime.

5. Calculate the Christoffel symbols for the Schwarzschild metric:

$$ds^2 = -\left(1 - \frac{2M}{r}\right) dt^2 + \left(1 - \frac{2M}{r}\right)^{-1} dr^2 + r^2 d\Omega^2,$$

focusing on the t and r components.

6. Describe the implications of the Hawking radiation phenomenon in the context of field quantization in curved spacetimes.

Answers 1

1. **Solution:**

   The action S is given as

   $$S = \int d^4x \sqrt{-g} \left(\frac{1}{2} g^{\mu\nu} \nabla_\mu \phi \nabla_\nu \phi - \frac{1}{2} m^2 \phi^2\right).$$

   Under a coordinate transformation $x^\mu \to x'^\mu$, the volume element d^4x changes as $d^4x' = \left|\frac{\partial x'}{\partial x}\right| d^4x$, where $\left|\frac{\partial x'}{\partial x}\right|$ is the determinant of the Jacobian of the transformation. The metric determinant g transforms as $g' = g \left|\frac{\partial x}{\partial x'}\right|^2$. Hence, $\sqrt{-g}\, d^4x$ is invariant.

   The covariant derivatives $\nabla_\mu \phi$ transform as tensors of rank 1, while $g^{\mu\nu}$ transforms as a tensor of rank 2, preserving the form of the action under coordinate transformations. Therefore, the action S is invariant under coordinate transformations.

2. **Solution:**

   The covariant derivative of a vector field V^ν is given by

   $$\nabla_\mu V^\nu = \partial_\mu V^\nu + \Gamma^\nu_{\mu\sigma} V^\sigma.$$

   The Christoffel symbols $\Gamma^\lambda_{\mu\nu}$ are defined as

   $$\Gamma^\lambda_{\mu\nu} = \frac{1}{2} g^{\lambda\sigma} \left(\partial_\mu g_{\sigma\nu} + \partial_\nu g_{\sigma\mu} - \partial_\sigma g_{\mu\nu}\right).$$

   This expression ensures that the covariant derivative of a vector is a tensor, maintaining its form under coordinate transformations.

3. **Solution:**

The mode decomposition of scalar fields is given by

$$\phi(x) = \sum_i \left(a_i u_i(x) + a_i^\dagger u_i^*(x) \right).$$

The field operators a_i and $a_i^\dagger$ satisfy the commutation relations:

$$[a_i, a_j^\dagger] = \delta_{ij}, \quad [a_i, a_j] = [a_i^\dagger, a_j^\dagger] = 0.$$

These relations arise from the orthogonality and completeness of the mode functions $u_i(x)$ in the chosen basis adapted to the geometry of the spacetime.

4. **Solution:**

In Minkowski spacetime, a global timelike Killing vector field exists, leading to a natural definition of positive frequency modes and hence a unique vacuum state. In curved spacetimes, such a global timelike Killing vector may not exist, making the concept of particles and the vacuum state coordinate-dependent. This ambiguity can lead to different observers perceiving different particle content in the same quantum field, exemplified by the Unruh effect.

5. **Solution:**

For the Schwarzschild metric:

$$ds^2 = - \left(1 - \frac{2M}{r} \right) dt^2 + \left(1 - \frac{2M}{r} \right)^{-1} dr^2 + r^2 d\Omega^2,$$

calculate Γ_{tt}^t and Γ_{rr}^r:

- $\Gamma_{tt}^t = \frac{1}{2} \left(-\partial_r \left(1 - \frac{2M}{r} \right) \right) = \frac{M}{r^2} \left(1 - \frac{2M}{r} \right)^{-1}.$
- $\Gamma_{rr}^r = \frac{1}{2} \partial_r \left(\left(1 - \frac{2M}{r} \right)^{-1} \right) = \frac{M}{r^2} \left(1 - \frac{2M}{r} \right)^{-1}.$

These symbols are crucial for computing geodesics and understanding curvature effects in the Schwarzschild geometry.

6. **Solution:**

Hawking radiation arises in the context of black holes as a consequence of particle creation in curved spacetime. Due to the event horizon, there is an asymmetry in modes that allows for thermal radiation to be emitted by the black hole. This effect challenges the classical view that nothing can escape a black hole and has profound implications for the preservation of information in quantum mechanics. The phenomenon is analyzed using field quantization techniques in the Schwarzschild spacetime, reflecting the effects of intense gravitational fields on quantum fields.

Practice Problems 2

1. Consider a scalar field ϕ in a curved spacetime with a metric tensor $g_{\mu\nu}$. Calculate the action S for the field, given by

$$S = \int d^4 x \sqrt{-g} \left(\frac{1}{2} g^{\mu\nu} \nabla_\mu \phi \nabla_\nu \phi - \frac{1}{2} m^2 \phi^2 \right).$$

2. For a vector field V^ν in a curved spacetime, compute the covariant derivative $\nabla_\mu V^\nu$, using the Christoffel symbols $\Gamma^\lambda_{\mu\nu}$.

$$\Gamma^\lambda_{\mu\nu} = \frac{1}{2}g^{\lambda\sigma}\left(\partial_\mu g_{\sigma\nu} + \partial_\nu g_{\sigma\mu} - \partial_\sigma g_{\mu\nu}\right).$$

3. Decompose the scalar field $\phi(x)$ into mode functions assuming a curved spacetime geometry.

$$\phi(x) = \sum_i \left(a_i u_i(x) + a_i^\dagger u_i^*(x)\right),$$

and provide the commutation relations for the creation and annihilation operators.

4. In the Schwarzschild metric, calculate the effect of curvature on the field equations for a scalar field by using the metric

$$ds^2 = -\left(1 - \frac{2M}{r}\right)dt^2 + \left(1 - \frac{2M}{r}\right)^{-1}dr^2 + r^2 d\Omega^2.$$

5. Analyze the vacuum state in a curved spacetime and explain how it may lead to phenomena such as Hawking radiation.

6. For a Friedmann-Lemaître-Robertson-Walker (FLRW) universe, represented by the metric

$$ds^2 = -dt^2 + a(t)^2 \left(\frac{dr^2}{1 - kr^2} + r^2 d\Omega^2 \right),$$

describe how quantum fields behave during the early universe and discuss the significance of the scale factor $a(t)$.

Answers 2

1. To find the action S for the scalar field ϕ in curved spacetime, notice that the expression involves tunneling the metric determinant $-g$, the inverse metric $g^{\mu\nu}$, and derivatives. Starting from the given expression:

$$S = \int d^4x \sqrt{-g} \left(\frac{1}{2} g^{\mu\nu} \nabla_\mu \phi \nabla_\nu \phi - \frac{1}{2} m^2 \phi^2 \right).$$

Use the covariant derivative $\nabla_\mu \phi = \partial_\mu \phi$ for scalars:

$$S = \int d^4x \sqrt{-g} \left(\frac{1}{2} g^{\mu\nu} \partial_\mu \phi \partial_\nu \phi - \frac{1}{2} m^2 \phi^2 \right).$$

This action emphasizes the dynamic nature of ϕ in curved spacetime by incorporating effects of curvature through covariant derivatives and metric integration.

2. To compute $\nabla_\mu V^\nu$, employ the definition of the covariant derivative for vector fields:

$$\nabla_\mu V^\nu = \partial_\mu V^\nu + \Gamma^\nu_{\mu\sigma} V^\sigma.$$

With the given Christopher symbols:

$$\Gamma^\lambda_{\mu\nu} = \frac{1}{2} g^{\lambda\sigma} \left(\partial_\mu g_{\sigma\nu} + \partial_\nu g_{\sigma\mu} - \partial_\sigma g_{\mu\nu} \right).$$

Use this expression for $\Gamma^\nu_{\mu\sigma}$ to determine $\nabla_\mu V^\nu$, which represents the curved spacetime adaptation of differentiation maintaining tensorial properties across coordinate changes.

3. For the mode decomposition of the field:

$$\phi(x) = \sum_i \left(a_i u_i(x) + a_i^\dagger u_i^*(x) \right),$$

where $u_i(x)$ and $u_i^*(x)$ are mode functions satisfying the field equations in the given spacetime. The commutation relations:

$$[a_i, a_j^\dagger] = \delta_{ij}, \quad [a_i, a_j] = [a_i^\dagger, a_j^\dagger] = 0,$$

ensure normalization and linear independence properties, important for developing a quantized field in curved backgrounds.

4. In Schwarzschild spacetime, the metric affects field equations by providing variable coefficients due to curvature. Specifically, the Schwarzschild metric:

$$ds^2 = -\left(1 - \frac{2M}{r}\right)dt^2 + \left(1 - \frac{2M}{r}\right)^{-1}dr^2 + r^2 d\Omega^2.$$

Use this to find how curvature influences the D'Alembertian operator $\Box_g \phi$, exhibiting effects like gravitational redshift of field propagation.

5. The vacuum state in curved spacetime, especially near event horizons, is nontrivial due to a lack of global inertial frames. This affects quantum fields, leading to Hawking radiation—a thermal flux emerging from black hole horizons. This phenomenon arises from quantum fluctuations in the vacuum and the misalignment of time translations between stationary observers.

6. In an FLRW universe, the metric:

$$ds^2 = -dt^2 + a(t)^2 \left(\frac{dr^2}{1 - kr^2} + r^2 d\Omega^2\right),$$

with the scale factor $a(t)$ describing cosmic expansion, defines the behavior of quantum fields. Initially proposed to model the Big Bang, fluctuations in this expanding universe permit studies of inflation, perturbations, and the resulting cosmic microwave background, highlighting the consilience of GR and QFT.

Practice Problems 3

1. Derive the variation of the action for a scalar field in curved spacetime, given by the action:

$$S = \int d^4x \sqrt{-g}\left(\frac{1}{2}g^{\mu\nu}\nabla_\mu\phi\nabla_\nu\phi - \frac{1}{2}m^2\phi^2\right).$$

2. Express the covariant derivative of a scalar field in a general coordinate system and simplify it.

$$\nabla_\mu \phi$$

3. Determine the Christoffel symbols for the Schwarzschild metric:

$$ds^2 = -\left(1 - \frac{2M}{r}\right)dt^2 + \left(1 - \frac{2M}{r}\right)^{-1}dr^2 + r^2 d\Omega^2.$$

4. Show how the mode decomposition of a scalar field appears in the context of a Friedmann-Lemaître-Robertson-Walker (FLRW) Universe.

$$ds^2 = -dt^2 + a(t)^2 \left(\frac{dr^2}{1 - kr^2} + r^2 d\Omega^2\right)$$

5. Explain the significance of Hawking radiation in the context of curved spacetime physics.

6. Discuss the concept of the Unruh effect and its physical implications in quantum field theory.

Answers 3

1. Derive the variation of the action for a scalar field in curved spacetime.

   Solution: To find the variation of the action, we compute the functional derivative with respect to ϕ and apply the principle of stationary action. The variation of the action is given by:

   $$\delta S = \int d^4 x \left[\frac{\delta \sqrt{-g}}{\delta \phi} \left(\frac{1}{2} g^{\mu\nu} \nabla_\mu \phi \nabla_\nu \phi - \frac{1}{2} m^2 \phi^2 \right) + \sqrt{-g} \delta \left(\frac{1}{2} g^{\mu\nu} \nabla_\mu \phi \nabla_\nu \phi - \frac{1}{2} m^2 \phi^2 \right) \right].$$

   Using integration by parts and neglecting boundary terms, the field equation derived is:

   $$\nabla_\mu \nabla^\mu \phi + m^2 \phi = 0,$$

   which is the Klein-Gordon equation in curved spacetime.

2. Express the covariant derivative of a scalar field in a general coordinate system.

   Solution: The covariant derivative of a scalar field ϕ in any coordinate system is:

   $$\nabla_\mu \phi = \partial_\mu \phi.$$

   For scalar fields, since they have no indices, the covariant derivative reduces to the ordinary partial derivative.

3. Determine the Christoffel symbols for the Schwarzschild metric.

   Solution: The Christoffel symbols, $\Gamma^\lambda_{\mu\nu}$, are computed using the metric components of the Schwarzschild metric. For the metric:

   $$ds^2 = - \left(1 - \frac{2M}{r} \right) dt^2 + \left(1 - \frac{2M}{r} \right)^{-1} dr^2 + r^2 d\Omega^2,$$

   the non-zero Christoffel symbols include:

   $$\Gamma^r_{tt} = \frac{M}{r^2} \left(1 - \frac{2M}{r} \right), \quad \Gamma^t_{tr} = \frac{M}{r^2 \left(1 - \frac{2M}{r} \right)},$$

   and others derived similarly by differentiating the metric components and applying the Christoffel definition:

   $$\Gamma^\lambda_{\mu\nu} = \frac{1}{2} g^{\lambda\sigma} \left(\partial_\mu g_{\sigma\nu} + \partial_\nu g_{\sigma\mu} - \partial_\sigma g_{\mu\nu} \right).$$

4. Show how the mode decomposition of a scalar field appears in a FLRW Universe.

   Solution: In a FLRW Universe with metric:

   $$ds^2 = -dt^2 + a(t)^2 \left(\frac{dr^2}{1 - kr^2} + r^2 d\Omega^2 \right),$$

   the scalar field $\phi(x)$ can be expressed as a mode decomposition:

   $$\phi(x, t) = \sum_k \left(a_k u_k(x, t) + a_k^\dagger u_k^*(x, t) \right),$$

   where the modes $u_k(x, t)$ are solutions to the field equations in the cosmological background, encapsulating the time evolution through $a(t)$.

5. Explain the significance of Hawking radiation.

   Solution: Hawking radiation is a quantum phenomenon where black holes emit radiation due to quantum effects near the event horizon. Its significance lies in its implication that black holes can lose mass and energy, eventually evaporating completely. This bridges quantum mechanics, thermodynamics, and general relativity, suggesting that information might not be conserved in black holes, known as the information paradox.

6. Discuss the concept of the Unruh effect.

   Solution: The Unruh effect describes the observation that an accelerating observer in vacuum perceives a thermal bath of particles, with temperature proportional to the acceleration. It illustrates that the concept of particles is observer-dependent in quantum field theory and highlights the non-trivial relationship between acceleration and temperature, implying that different observers can have divergent experiences of the concept of a vacuum.

Chapter 31

Supersymmetry and Super Hilbert Spaces

Introduction to Supersymmetric Quantum Field Theories

Supersymmetry is a theoretical framework in quantum field theory that posits a symmetry between bosons and fermions. Bosons have integer spins and obey Bose-Einstein statistics, whereas fermions have half-integer spins and obey Fermi-Dirac statistics. Supersymmetric theories introduce a new operator, the supercharge, which relates bosonic and fermionic states. This relationship is encapsulated in the supersymmetry algebra.

The basic supersymmetry algebra involves generators Q_α and $\bar{Q}^{\dot{\alpha}}$, which satisfy the anticommutation relations:

$$\{Q_\alpha, \bar{Q}^{\dot{\beta}}\} = 2\sigma^\mu_{\alpha\dot{\beta}} P_\mu, \tag{31.1}$$

where σ^μ are the Pauli matrices and P_μ represents the four-momentum operator. This algebra reflects the core principle of supersymmetry: every bosonic degree of freedom has a fermionic counterpart and vice versa.

Construction of Super Hilbert Spaces

A natural setting for supersymmetric quantum mechanics is the concept of a super Hilbert space. In a super Hilbert space, states can be thought of as combinations of both bosonic and fermionic components. These states are formalized using the language of **supervectors** and **superoperators**.

1 Supervectors and Superoperators

In the context of supersymmetry, a supervector Φ can be expressed as:

$$\Phi = \begin{pmatrix} \phi \\ \psi \end{pmatrix}, \tag{31.2}$$

where ϕ is a bosonic field and ψ is a fermionic field. The operations on these vectors require the definition of superoperators which act on super Hilbert spaces, maintaining the distinction between the bosonic and fermionic components.

The superoperators are matrices of the form:

$$\mathcal{O} = \begin{pmatrix} A & B \\ C & D \end{pmatrix}, \tag{31.3}$$

where A and D are operators acting within bosonic and fermionic subspaces respectively, while B and C mix these subspaces.

2   Inner Product and Norm in Super Hilbert Spaces

The inner product in a super Hilbert space is defined to respect the grading of the space. For two supervectors $\Phi_1 = \begin{pmatrix} \phi_1 \\ \psi_1 \end{pmatrix}$ and $\Phi_2 = \begin{pmatrix} \phi_2 \\ \psi_2 \end{pmatrix}$, the inner product is given by:

$$\langle \Phi_1, \Phi_2 \rangle = \langle \phi_1, \phi_2 \rangle + \langle \psi_1, \psi_2 \rangle, \tag{31.4}$$

where $\langle \cdot, \cdot \rangle$ are the standard inner products in bosonic and fermionic sectors, respectively. A supervector's norm is given by the square root of the inner product with itself:

$$\|\Phi\| = \sqrt{\langle \Phi, \Phi \rangle}. \tag{31.5}$$

Applications of Supersymmetry in Quantum Field Theory

Supersymmetry introduces elegant cancellations in quantum field theories, reducing the severity of ultraviolet divergences through the balancing of bosonic and fermionic loops. This principle ameliorates the hierarchy problem by stabilizing the Higgs boson mass through supersymmetric partners. Furthermore, the structure of supersymmetry naturally leads to the unification of gauge couplings at high energy scales, providing a compelling framework for searching for theories beyond the Standard Model.

Role of Super Hilbert Spaces in Supersymmetric Field Theories

Super Hilbert spaces facilitate the unification of bosonic and fermionic fields within a single framework. By structuring these spaces with respect to their supersymmetry algebra, computations in quantum field theory become more tractable, leveraging the algebraic properties of superspaces. They also provide a formal basis for the construction of supersymmetric versions of quantum mechanics, crucial for exploring phenomenological implications in particle physics.

Beyond symmetry aspects, super Hilbert spaces accommodate the algebraic structures necessary to understand spontaneous symmetry breaking and the emergence of supersymmetric vacua, integral to the realization of realistic physical models.

Practice Problems 1

1. Verify the anticommutation relation of the supersymmetry generators given by Equation 31.1 in terms of the Pauli matrices σ^μ and the four-momentum operator P_μ. 6cm

2. Determine the conditions under which a given operator in the form of $\mathcal{O} = \begin{pmatrix} A & B \\ C & D \end{pmatrix}$ acts unitarily on a super Hilbert space. 6cm

3. Show that the inner product defined in Section 2 is Hermitian and positive definite for supervectors Φ. 6cm

4. Given a supervector $\Phi = \begin{pmatrix} \phi \\ \psi \end{pmatrix}$, compute the norm using Equation 31.5 if $\phi = 3 + 4i$ and $\psi = 1 - i$. 6cm

5. Explore the cancellation of ultraviolet divergences in a supersymmetric quantum field theory with a simple loop diagram. 6cm

6. Discuss the role of spontaneous symmetry breaking in the context of super Hilbert spaces and supersymmetry, specifically its implications for the physical realization of supersymmetric theories. 6cm

Answers 1

1. Verify the anticommutation relation of the supersymmetry generators given by Equation 31.1:

   Solution: The anticommutation relation is given as:

   $$\{Q_\alpha, \bar{Q}^{\dot{\beta}}\} = 2\sigma^\mu_{\alpha\dot{\beta}} P_\mu$$

   To verify, we define Q_α and $\bar{Q}^{\dot{\beta}}$ to satisfy:

   - Q_α and $\bar{Q}^{\dot{\beta}}$ are assumed as fermionic operators.
   - The commutator $[\sigma^\mu, \sigma^\nu] \neq 0$.

   Considering these properties, we apply the definition of the Pauli matrices and highlight the requirement:

   - P_μ is the four-momentum operator obeying relativistic invariance.

   By substituting into the commutators and considering the algebraic properties of σ^μ, the anticommutation relation is thus verified by its structure ensuring the operator balance between fermionic and bosonic states.

2. Determine the conditions for $\mathcal{O}$ to be unitary:

   Solution: A superoperator $\mathcal{O}$ is unitary if:

   $$\mathcal{O}^\dagger \mathcal{O} = \mathbb{I}$$

   For $\mathcal{O} = \begin{pmatrix} A & B \\ C & D \end{pmatrix}$:

   $$\mathcal{O}^\dagger = \begin{pmatrix} A^\dagger & C^\dagger \\ B^\dagger & D^\dagger \end{pmatrix}$$

   Thus, $\mathcal{O}^\dagger \mathcal{O}$ results in:

   $$\begin{pmatrix} A^\dagger A + B^\dagger C & A^\dagger B + B^\dagger D \\ C^\dagger A + D^\dagger C & C^\dagger B + D^\dagger D \end{pmatrix} = \begin{pmatrix} \mathbb{I} & 0 \\ 0 & \mathbb{I} \end{pmatrix}$$

   Verifying each element, unitarity holds if:
   - $A^\dagger A + B^\dagger C = \mathbb{I}$ and similar for other diagonal terms.
   - Off-diagonal terms are zero.

3. Show Hermitian and positive definiteness of the inner product:

   Solution: Using:

   $$\langle \Phi_1, \Phi_2 \rangle = \langle \phi_1, \phi_2 \rangle + \langle \psi_1, \psi_2 \rangle$$

   Hermitian:

   $$\langle \Phi_1, \Phi_2 \rangle = \overline{\langle \Phi_2, \Phi_1 \rangle}$$

   Verifies by symmetry properties of complex conjugates.

   Positive definiteness: For $\Phi = \begin{pmatrix} \phi \\ \psi \end{pmatrix}$:

   $$\langle \Phi, \Phi \rangle \geq 0$$

   Equality only for zero vector due to properties of $\langle \phi, \phi \rangle$.

4. Compute the norm for $\Phi = \begin{pmatrix} \phi \\ \psi \end{pmatrix} = \begin{pmatrix} 3 + 4i \\ 1 - i \end{pmatrix}$:

   Solution: Calculate each component's magnitude:

   $$\langle \phi, \phi \rangle = (3 + 4i)(3 - 4i) = 9 + 16 = 25$$

   $$\langle \psi, \psi \rangle = (1 - i)(1 + i) = 1 + 1 = 2$$

   Norm:

   $$\|\Phi\| = \sqrt{25 + 2} = \sqrt{27} = 3\sqrt{3}$$

5. Cancellation of ultraviolet divergences:

   Solution: Supersymmetry aligns bosonic and fermionic contributions:

   - Each boson loop divergence $\sim \Lambda^2$ is offset by a fermion loop $\sim -\Lambda^2$.
   - For symmetry-preserving regularization, they accurately cancel beyond tree levels, maintaining finite corrections.

6. Discuss spontaneous symmetry breaking:

   Solution: With spontaneous symmetry breaking in supersymmetric contexts:

   - Super Hilbert spaces include both bosonic and fermionic vacua.
   - After breaking, mass degeneracy among particles alters, observing super potential roles.
   - Realizes masses in MSSM (Minimal Supersymmetric Standard Model), predicting new particles, confirming with observables distinct from unbroken counterparts.

Practice Problems 2

1. Verify the supersymmetry algebra by proving the anticommutation relation:

$$\{Q_\alpha, \bar{Q}^{\dot\beta}\} = 2\sigma^\mu_{\alpha\dot\beta} P_\mu.$$

2. Construct a supervector from a given bosonic field $\phi(x)$ and a fermionic field $\psi(x)$, and express it in the form:

$$\Phi = \begin{pmatrix} \phi \\ \psi \end{pmatrix}.$$

3. Given the superoperator $\mathcal{O} = \begin{pmatrix} A & B \\ C & D \end{pmatrix}$, show that this operator maintains the supersymmetry transformations in a super Hilbert space.

4. Derive the inner product for two supervectors Φ_1 and Φ_2, given by:

$$\Phi_1 = \begin{pmatrix} \phi_1 \\ \psi_1 \end{pmatrix}, \quad \Phi_2 = \begin{pmatrix} \phi_2 \\ \psi_2 \end{pmatrix}.$$

5. Explain the role of super Hilbert spaces in addressing the hierarchy problem in supersymmetric models.

6. Discuss how the concept of super Hilbert spaces facilitates the study of gauge coupling unification in supersymmetric theories.

Answers 2

1. **Solution:**

   To verify the supersymmetry algebra, we begin by recalling the general anticommutation relation:

   $$\{Q_\alpha, \bar{Q}^{\dot\beta}\} = 2\sigma^\mu_{\alpha\dot\beta} P_\mu.$$

   This indicates that the commutator of the supercharges Q_α and $\bar{Q}^{\dot\beta}$ relates to the four-momentum operator P_μ. The proof involves demonstrating this relation through calculations using Grassmann variables and the properties of the Pauli matrices σ^μ. This confirms that acting with these operators transforms bosonic to fermionic states and vice versa, respecting the underlying symmetry of the theory.

2. **Solution:**

305

To construct a supervector, we start by identifying the bosonic field $\phi(x)$ and the fermionic field $\psi(x)$. These fields represent the two components of our supervector. We then express it as:

$$\Phi = \begin{pmatrix} \phi \\ \psi \end{pmatrix}.$$

Here, $\phi(x)$ is the top component representing the bosonic part, while $\psi(x)$ is the bottom component for the fermionic part. This structure respects the grading required in super Hilbert spaces.

3. **Solution:**

The given superoperator $\mathcal{O}$ acts on supervectors in a super Hilbert space, maintaining the symmetry between bosonic and fermionic components. To show that this operator preserves supersymmetry transformations, consider the matrix form:

$$\mathcal{O} = \begin{pmatrix} A & B \\ C & D \end{pmatrix}.$$

The operator's structure ensures that bosonic operators A and D transform their respective sectors, while B and C are responsible for interactions between the bosonic and fermionic parts. The preservation of supersymmetry transformations is evident from ensuring these matrices satisfy the commutation and anticommutation relations which stem from the supersymmetry algebra.

4. **Solution:**

The inner product $\langle \Phi_1, \Phi_2 \rangle$ in a super Hilbert space connects the components of two supervectors:

$$\langle \Phi_1, \Phi_2 \rangle = \langle \phi_1, \phi_2 \rangle + \langle \psi_1, \psi_2 \rangle.$$

Here, $\langle \phi_1, \phi_2 \rangle$ and $\langle \psi_1, \psi_2 \rangle$ represent inner products in their respective bosonic and fermionic spaces. This definition respects the grading of bosonic and fermionic components, ensuring consistency with the conventional inner product rules.

5. **Solution:**

Super Hilbert spaces play a crucial role in supersymmetric models addressing the hierarchy problem. The hierarchy problem involves the large disparity between the weak force and gravitational forces, particularly regarding the Higgs boson mass. Supersymmetry proposes partner particles for each Standard Model particle, introducing cancellations in quantum corrections. Super Hilbert spaces, by providing a structure that can describe both bosons and fermions, facilitate the formulation and computation of these cancellations, thereby stabilizing mass hierarchies.

6. **Solution:**

In the pursuit of gauge coupling unification, super Hilbert spaces serve as a vital framework. Supersymmetry implies that the gauge couplings—parameters that determine the strength of forces—converge at high-energy scales. The unification is natural in supersymmetric theories due to the specific energy scale dependence of these couplings. Super Hilbert spaces define the quantum states and operators in a unified manner, making the calculations involving these couplings more tractable and allowing physicists to elucidate the conditions under which unification successfully occurs.

Practice Problems 3

1. Verify that the supersymmetry algebra given by the anticommutation relations in Equation 31.1 satisfies the Jacobi identity.

2. Demonstrate the transformation of a bosonic field ϕ and a fermionic field ψ under a supersymmetry transformation using the supercharges Q_α and $\bar{Q}^{\dot{\alpha}}$.

3. Prove that the norm defined in a super Hilbert space (Equation 31.5) remains invariant under unitary transformations.

4. Show that the operator matrix $\mathcal{O}$, as defined in Equation 31.3, preserves the grading of a supervector when it acts on it.

5. Discuss the implications of supersymmetry in reducing the degree of ultraviolet divergences in quantum field theories, specifically referencing loop corrections.

6. Explain the role of super Hilbert spaces in overcoming challenges associated with the hierarchy problem in quantum field theories.

Answers 3

1. Verify that the supersymmetry algebra given by the anticommutation relations in Equation 31.1 satisfies the Jacobi identity.
   Solution:
   The Jacobi identity for operators A, B, and C is given by:

   $$\{A, \{B, C\}\} + \{B, \{C, A\}\} + \{C, \{A, B\}\} = 0.$$

   For the supersymmetry algebra, consider the operators Q_α, $\bar{Q}^{\dot{\alpha}}$, and P_μ. We need to check whether the above condition holds for these operators. Using the relations:

   $$\{Q_\alpha, \bar{Q}^{\dot{\beta}}\} = 2\sigma^\mu_{\alpha\dot{\beta}} P_\mu,$$

   and considering the trivial anti-commutation among four-momentum operators:

   $$\{P_\mu, P_\nu\} = 0,$$

   the Jacobi identity becomes trivial for any cyclic permutations, affirming that it holds.

2. Demonstrate the transformation of a bosonic field ϕ and a fermionic field ψ under a supersymmetry transformation using the supercharges Q_α and $\bar{Q}^{\dot{\alpha}}$.
   Solution:
   A supersymmetry transformation can be represented as:

   $$\delta_\epsilon \phi = \epsilon^\alpha Q_\alpha \phi, \quad \delta_\epsilon \psi = \epsilon^\alpha Q_\alpha \psi + \bar{\epsilon}_{\dot{\alpha}} \bar{Q}^{\dot{\alpha}} \phi.$$

   For a bosonic field ϕ, the transformation mainly results in a shift proportional to a fermionic field due to its pairing:

   $$\delta_\epsilon \psi \equiv \epsilon^\alpha \partial_\mu \phi \rightarrow \text{fermionic component.}$$

3. Prove that the norm defined in a super Hilbert space (Equation 31.5) remains invariant under unitary transformations.
   Solution:
   Consider a unitary transformation $\mathcal{U}$ on a supervector Φ,

   $$\Phi' = \mathcal{U}\Phi.$$

   The transformed norm is:
   $$\|\Phi'\| = \sqrt{\langle \Phi', \Phi' \rangle}.$$
   Since $\mathcal{U}$ is unitary, $\langle \mathcal{U}\Phi, \mathcal{U}\Phi \rangle = \langle \Phi, \Phi \rangle$, thus,

   $$\|\Phi'\| = \|\Phi\|.$$

4. Show that the operator matrix $\mathcal{O}$, as defined in Equation 31.3, preserves the grading of a supervector when it acts on it.
   Solution:
   An operator matrix $\mathcal{O} = \begin{pmatrix} A & B \\ C & D \end{pmatrix}$ acts on a supervector $\Phi = \begin{pmatrix} \phi \\ \psi \end{pmatrix}$. When applied,

   $$\mathcal{O}\Phi = \begin{pmatrix} A\phi + B\psi \\ C\phi + D\psi \end{pmatrix}.$$

   A, D maintain the bosonic/fermionic structure, while B, C change bosonic to fermionic and vice versa but the total grading is conserved, demonstrating preservation of grading.

5. Discuss the implications of supersymmetry in reducing the degree of ultraviolet divergences in quantum field theories, specifically referencing loop corrections.

Solution:

Supersymmetry provides additional canceling interactions in loop diagrams, reducing the number of UV divergences. In a one-loop correction, paired bosonic and fermionic loops can contribute opposite signs due to the statistics (Bose-Einstein vs. Fermi-Dirac), leading to partial or complete cancellations. This property is particularly advantageous in preventing large corrections to scalar masses, like the Higgs, addressing the hierarchy problem.

6. Explain the role of super Hilbert spaces in overcoming challenges associated with the hierarchy problem in quantum field theories.

Solution:

The hierarchy problem involves large quantum corrections to the Higgs mass, potentially destabilizing it. Super Hilbert spaces provide a framework where each particle's fermionic and bosonic counterparts balance these corrections. The unification within a super Hilbert space ensures that divergent contributions in supersymmetric partners cancel each other, thereby maintaining a stable, small Higgs mass. This cancellation is automatic in a supersymmetrically balanced system, pointing towards the naturalness of the Higgs without fine-tuning.

Chapter 32

Noncommutative Geometry in Quantum Field Theory

Introduction to Noncommutative Geometry

Noncommutative geometry extends the concepts of geometry to settings where the algebra of functions on a space does not commute. A fundamental example of this is the algebra of observables in quantum mechanics, where operators representing physical quantities do not necessarily commute. The central idea of noncommutative geometry is to replace the notion of a point in a space with algebraic objects, typically noncommutative algebras, which suggest a "quantum" space.

Consider C^*-algebras as a generalization of space: traditional geometric spaces are described by C-algebras of continuous functions with pointwise multiplication and complex conjugation. In noncommutative geometry, these are extended to noncommutative C^*-algebras.

Algebraic Framework

The algebraic framework involves the study of noncommutative algebras, typically Banach algebras that generalize the idea of function spaces. A C^*-algebra $\mathcal{A}$ is defined through the completenes of the space with respect to a norm $\|\cdot\|$ satisfying:

$$\|ab\| \leq \|a\|\|b\|, \tag{32.1}$$

for all $a, b \in \mathcal{A}$. Moreover, the *-operation must satisfy:

$$\|a^*a\| = \|a\|^2. \tag{32.2}$$

The fundamental nature of noncommutative geometry comes from the failure of multiplication in $\mathcal{A}$ to be commutative, meaning:

$$ab \neq ba. \tag{32.3}$$

1 Spectral Triples

A spectral triple $(\mathcal{A}, \mathcal{H}, D)$ provides a framework to re-interpret geometry in algebraic language. Here, $\mathcal{A}$ is a C^*-algebra acting on a Hilbert space $\mathcal{H}$. The operator D, usually a self-adjoint unbounded operator, generalizes the Dirac operator from classical geometry. It plays a role analogous to the inverse of a metric in Riemannian geometry.

The following properties must be satisfied:

1. The commutator $[D, a]$ is bounded for each $a \in \mathcal{A}$. 2. $(1 + D^2)^{-1}$ is compact. 3. D has discrete spectrum.

Noncommutative Spaces in Quantum Field Theory

In quantum field theory (QFT), spacetime is treated as a continuum. Noncommutative geometry replaces points in spacetime with noncommutative algebraic objects, providing a natural framework for formulating QFT on these new 'quantum spaces'.

1 Moyal Product

The Moyal product is an example of a noncommutative product useful in the formulation of noncommutative field theories. For functions f and g on $\mathbb{R}^n$, their Moyal product $f \star g$ is given by:

$$(f \star g)(x) = f(x) \exp\left(\frac{i}{2}\theta^{ij}\overleftarrow{\partial_i}\overrightarrow{\partial_j}\right) g(x), \tag{32.4}$$

where θ^{ij} is an antisymmetric matrix determining the noncommutativity of the coordinates, and $\overleftarrow{\partial_i}$, $\overrightarrow{\partial_j}$ denote derivatives acting to the left and right, respectively.

2 Field Theory on Noncommutative Spaces

Quantum field theories on noncommutative spaces require modifications to standard field theory frameworks to accommodate the noncommutative product. The action for a scalar field ϕ becomes:

$$S = \int d^n x \left(\frac{1}{2}\partial_i\phi \star \partial^i\phi - \frac{\mu^2}{2}\phi \star \phi - \frac{\lambda}{4}\phi \star \phi \star \phi \star \phi\right). \tag{32.5}$$

This action embodies the replacement of pointwise multiplication with the Moyal star product. Such actions define the behavior of fields on noncommutative spacetimes and affect both the propagators and interactions in perturbative expansions.

Applications and Implications

The application of noncommutative geometry in quantum field theory impacts both the infrared and ultraviolet behavior of field theories. One of the remarkable implications is the potential resolution of the ultraviolet-divergence problem by incorporating a minimal length scale naturally within the geometry. Additionally, noncommutative geometry offers a route to integrate gravity into quantum frameworks, suggesting modifications to general relativity on quantum scales.

Noncommutative field theories allow exploration of phenomena at scales approaching the Planck scale, where quantum gravity effects become significant. They provide not only a mathematical structure for such extensions but also a conceptual tool for investigating the limits of current physical theories.

Practice Problems 1

1. Describe how noncommutative geometry generalizes classical geometry using the framework of C^*-algebras. What implications does this have for the nature of space?

2. Explain the role of spectral triples in noncommutative geometry. What are the components of a spectral triple and their geometric significance?

3. Demonstrate how the Moyal product introduces noncommutativity in the algebra of functions. Calculate $(x \star y)(z)$ for simple functions $f(x) = x$ and $g(x) = y$.

4. Define and examine the general conditions under which an operator in a spectral triple has a discrete spectrum. Why is this requirement important?

5. Analyze the impact of noncommutative geometry on quantum field theories. What modifications occur in the field equations and interaction terms?

6. Discuss the potential resolutions to the ultraviolet divergence problem that noncommutative geometry might offer. How does this relate to the introduction of a minimal length scale?

Answers 1

1. **Solution:** Noncommutative geometry generalizes classical geometry by replacing the algebra of commutative functions $C(X)$ on a space X with a noncommutative C^*-algebra $\mathcal{A}$. Traditionally, points in a space are described by evaluating functions at a point. In noncommutative geometry, these points are ill-defined, and geometry is described in terms of algebraic relations within $\mathcal{A}$. This implies that the concept of a space can be viewed through the properties and structures of its algebra, transforming our understanding of space into one dominated by algebraic, rather than purely topological, properties.

2. **Solution:** A spectral triple $(\mathcal{A}, \mathcal{H}, D)$ consists of a C^*-algebra $\mathcal{A}$, a Hilbert space $\mathcal{H}$, and a self-adjoint operator D. The algebra $\mathcal{A}$ acts on $\mathcal{H}$ as bounded operators, and D serves as a generalization of the Dirac operator, connecting the algebraic structure to geometric and analytic properties. This setup allows us to describe geometric notions like distance and dimensionality algebraically, with D playing a role similar to that of metric tensors in Riemannian geometry.

3. **Solution:** The Moyal product $(f \star g)(x) = f(x) \exp\left(\frac{i}{2}\theta^{ij}\overleftarrow{\partial_i}\overrightarrow{\partial_j}\right) g(x)$ introduces noncommutativity by using derivatives and a deformation parameter θ. For simple linear functions $f(x) = x$ and $g(x) = y$, assuming θ^{ij} is such that:
$$(x \star y)(z) = x \cdot \left(\exp\left(\frac{i}{2}\theta^{ij}\overleftarrow{\partial_i}\overrightarrow{\partial_j}\right)\right) \cdot y$$

becomes
$$xy + \frac{i}{2}\theta^{ij}[x, y] = xy + \frac{i}{2}\theta^{xy} \neq yx$$

illustrating noncommutativity.

4. **Solution:** For an operator D in a spectral triple to have a discrete spectrum, it should be self-adjoint and compact resolvent, meaning $(1 + D^2)^{-1}$ is compact. This requirement ensures that the noncommutative space adheres to certain compactness and continuity properties analogous to classical spaces. Discrete spectrum implies that the operator's action results in countable or isolated set outcomes, akin to quantized energy levels in quantum systems, providing a bridge between discrete observations (spectrum) and continuous operator theory.

5. **Solution:** In quantum field theories, replacing pointwise multiplication with noncommutative products such as the Moyal product alters field equations. For instance, the action for a scalar field becomes dependent on noncommutative products, modifying interaction terms like $\phi \star \phi$ instead of ϕ^2. These modifications impact kinetic and potential terms, leading to new dispersion relations and interaction dynamics, requiring novel renormalization techniques to handle infinities or divergences inherent in quantum fields.

6. **Solution:** Noncommutative geometry proposes a potential solution to the ultraviolet divergence problem by limiting resolution to a minimal length scale, naturally mitigating high-energy infinities. This length scale emerges from the noncommutative structure, embedding a fundamental discreteness within spacetime descriptions. This inherently smoothes out divergences traditionally encountered in continuous classical spacetime models, providing quantum-consistent spacetime frameworks that align with observed phenomenology at quantum scales.

Practice Problems 2

1. Verify that the commutator $[D, a] = Da - aD$ is bounded for an element $a \in \mathcal{A}$ in a spectral triple $(\mathcal{A}, \mathcal{H}, D)$.

2. Explain why $(1 + D^2)^{-1}$ is compact in the setting of a spectral triple $(\mathcal{A}, \mathcal{H}, D)$.

3. Show that the operator D has a discrete spectrum in the context of spectral triples in noncommutative geometry.

4. Prove that the Moyal product satisfies associativity for functions f, g, h.

5. Derive the modified action for a quantum field theory using the Moyal product for a scalar field ϕ, as presented in the chapter.

6. Discuss the significance of noncommutative geometry in resolving the ultraviolet-divergence problem in quantum field theory.

Answers 2

1. Verify that the commutator $[D, a] = Da - aD$ is bounded for an element $a \in \mathcal{A}$ in a spectral triple $(\mathcal{A}, \mathcal{H}, D)$.

   Solution: The commutator $[D, a]$ is bounded if there exists some constant C such that for all $\psi \in \mathcal{H}$, the inequality $\|[D, a]\psi\| \leq C\|\psi\|$ holds. The property of the spectral triple requires that $[D, a]$ is bounded for each $a \in \mathcal{A}$. This is a part of the definition of spectral triples, ensuring that geometric properties of spaces can be captured by noncommutative algebras.

2. Explain why $(1 + D^2)^{-1}$ is compact in the setting of a spectral triple $(\mathcal{A}, \mathcal{H}, D)$.

   Solution: The operator $(1 + D^2)^{-1}$ is compact if it can be approximated by finite rank operators in the operator norm. In the spectral triple context, D has a discrete spectrum allowing its eigenvectors to form an orthonormal basis. The operator $(1 + D^2)^{-1}$ thus acts like a smoothing operator which makes it compact because it implies that large eigenvalues are suppressed, making it feasible to approximate by operators with finite-dimensional range.

3. Show that the operator D has a discrete spectrum in the context of spectral triples in noncommutative geometry.

   Solution: The operator D in a spectral triple is typically assumed to be a self-adjoint operator on the Hilbert space $\mathcal{H}$. The discreteness of the spectrum means D has eigenvalues that are isolated points of its spectrum. This is often part of the structure of the spectral triple, where the algebraic-functional structure imposed ensures that D has discrete spectrum. This property allows for a well-defined notion of metric in the noncommutative setting via the generalized Dirac operator.

4. Prove that the Moyal product satisfies associativity for functions f, g, h.

   Solution: To prove associativity, we show $(f \star g) \star h = f \star (g \star h)$. Consider: $(f \star g) \star h = (f \exp(i\theta^{ij}\partial_i \otimes \partial_j)g) \star h$. Applying the Moyal product again, we expand: $= (f \star g) \exp(i\theta^{ij}\partial_i \otimes \partial_j)h$.

315

The exponential properties keep the ordering consistent, which preserves associativity. When considering formal expansions of the exponential in terms of its series, convergence ensures equivalence and associativity in practical terms.

5. Derive the modified action for a quantum field theory using the Moyal product for a scalar field ϕ, as presented in the chapter.

   Solution: The action S using the Moyal product for a scalar field ϕ is:

   $$S = \int d^n x \left(\frac{1}{2} \partial_i \phi \star \partial^i \phi - \frac{\mu^2}{2} \phi \star \phi - \frac{\lambda}{4} \phi \star \phi \star \phi \star \phi \right).$$

   Here, standard multiplication is replaced by the Moyal star product. This replacement modifies interactions and kinematic terms which impacts the entire field theory's behavior.

6. Discuss the significance of noncommutative geometry in resolving the ultraviolet-divergence problem in quantum field theory.

   Solution: Noncommutative geometry changes the structure of spacetime itself, introducing a natural UV cutoff by smearing points over areas defined by noncommutative parameters. The UV divergences in standard QFT often arise from treating spacetime as continuous and having interactions that concentrate infinitely small. Noncommutative settings inherently prevent such pathologies by altering how fields interact and propagate, particularly at very short distances associated with such divergences. This potential to regulate infinities signifies a profound method to reconcile QFT with quantum gravity scales.

Practice Problems 3

1. Explain why the Moyal product $(f \star g)(x)$ defined as:

   $$(f \star g)(x) = f(x) \exp \left(\frac{i}{2} \theta^{ij} \overleftarrow{\partial_i} \overrightarrow{\partial_j} \right) g(x)$$

   is noncommutative.

2. Prove that for a noncommutative C^*-algebra $\mathcal{A}$, the *-operation satisfies $\|a^* a\| = \|a\|^2$.

3. Describe the role of the Dirac operator D in the spectral triple $(\mathcal{A}, \mathcal{H}, D)$.

4. Derive the modified action in noncommutative field theory for a scalar field using the Moyal star product.

5. Discuss the implications of noncommutative geometry on ultraviolet divergences in quantum field theory.

6. Explain how noncommutative geometry may offer a framework for integrating gravity into quantum field theories.

Answers 3

1. **Explain why the Moyal product $(f \star g)(x)$ is noncommutative.**
   Solution:

The Moyal product between two functions f and g involves the operation:

$$(f \star g)(x) = f(x) \exp\left(\frac{i}{2}\theta^{ij}\overleftarrow{\partial_i}\overrightarrow{\partial_j}\right) g(x).$$

Here, the $\exp\left(\frac{i}{2}\theta^{ij}\overleftarrow{\partial_i}\overrightarrow{\partial_j}\right)$ operator acts on the partial derivatives of the functions f and g, introducing a noncommutative feature through the parameter θ^{ij}, which is antisymmetric. The antisymmetry leads to a scenario where generally $f \star g \neq g \star f$, hence making the product noncommutative.

2. **Prove that for a noncommutative C^*-algebra $\mathcal{A}$, the *-operation satisfies $\|a^*a\| = \|a\|^2$.**

Solution:
A C^*-algebra is equipped with a norm $\|\cdot\|$ and a *-operation (conjugate transpose for matrices) such that:

$$\|a^*a\| = \|a\|^2 \quad \text{for all} \quad a \in \mathcal{A}.$$

The proof is based on the axioms of a C^*-algebra. A key property is:

$$\|a^*\| = \|a\|,$$

meaning the norm is preserved under the *-operation. Now for any element a, consider:

$$\|a^*a\| \leq \|a^*\|\|a\| = \|a\|^2.$$

Conversely, if $\|a\|^2$ is achieved, it implies the given property. This forms the cornerstone of C^*-algebra definitions, confirming that indeed $\|a^*a\| = \|a\|^2$.

3. **Describe the role of the Dirac operator D in the spectral triple $(\mathcal{A}, \mathcal{H}, D)$.**

Solution:
In a spectral triple $(\mathcal{A}, \mathcal{H}, D)$:
- $\mathcal{A}$ is a C^*-algebra acting on a Hilbert space $\mathcal{H}$.
- D is typically a self-adjoint unbounded operator that serves as a generalization of the Dirac operator in a Riemannian manifold.

The role of the Dirac operator is multifaceted: 1. **Geometry Interpretation**: It plays a similar role to a metric in classical geometry, helping define distances. 2. **Spectral Data**: The spectrum of D encodes geometric information about the space, potentially revealing insights about the underlying "quantum" geometry. 3. **Commutator Condition**: The condition $[D, a]$ is bounded for $a \in \mathcal{A}$ ensures that D captures the geometric nature of algebraically defined spaces.

4. **Derive the modified action in noncommutative field theory for a scalar field using the Moyal star product.**

Solution:
The action S for a scalar field ϕ in a noncommutative field theory is modified using the Moyal product as follows:

$$S = \int d^n x \left(\frac{1}{2}(\partial_i\phi \star \partial^i\phi) - \frac{\mu^2}{2}(\phi \star \phi) - \frac{\lambda}{4}(\phi \star \phi \star \phi \star \phi)\right).$$

Each product now involves the Moyal star product:

$$(f \star g)(x) = f(x) \exp\left(\frac{i}{2}\theta^{ij}\overleftarrow{\partial_i}\overrightarrow{\partial_j}\right) g(x).$$

This structure inherently alters the formulation to account for noncommutative effects, affecting interaction terms and changing the computational framework of perturbative expansions.

5. **Discuss the implications of noncommutative geometry on ultraviolet divergences in quantum field theory.**

   Solution:

   Noncommutative geometry has significant implications for handling ultraviolet (UV) divergences in quantum field theory:

   - **Minimal Length Scale**: Introducing noncommutative structures suggests a minimal length scale, effectively regulating field interactions.
   - **New Feynman Diagrams**: Calculations must accommodate noncommutative parameters, leading to altered convergence properties in perturbation theory.
   - **Smooth UV Behavior**: The inherent structure can soften or remove divergence issues that plague traditional quantum field theories.
   - **Modified Renormalization**: Noncommutative field theories often require novel renormalization techniques due to altered propagators and interactions.

6. **Explain how noncommutative geometry may offer a framework for integrating gravity into quantum field theories.**

   Solution:

   Noncommutative geometry can potentially provide a powerful framework for integrating gravity into quantum field theories as follows:

   - **Geometrization of Quantum Gravity**: By treating spacetime as a quantum entity with a noncommutative algebra structure, traditional incompatibilities with gravity are bypassed.
   - **Induced Gravitational Theories**: Noncommutative spaces generate curvature automatically, suggesting a gravitational background arising naturally.
   - **Planck Scale Effects**: At such extreme scales, where quantum gravity is significant, noncommutative models inherently handle quantum fluctuations.
   - **Conceptual Unification**: By providing a consistent algebraic and geometric formulation, noncommutative geometry offers a unified approach to blending quantum mechanics with geometric spacetime frameworks.

Chapter 33

Microlocal Analysis in Quantum Field Theory

Introduction to Microlocal Analysis

Microlocal analysis provides a refined approach to the study of partial differential equations (PDEs) by focusing on their behavior at a micro scale. This technique enhances the traditional Fourier analysis by combining it with the geometry of phase space, utilizing tools such as wavefront sets to understand singularities.

Wavefront Sets and Singularities

Wavefront sets offer insight into the singularities of distributions, a cornerstone concept in microlocal analysis. For a distribution u, the wavefront set $\mathrm{WF}(u)$ is defined to capture the joint information on the location and directional elements of singularities.

1   Definition and Properties

The wavefront set $\mathrm{WF}(u)$ is a subset of the cotangent space $T^*X \setminus 0$, comprising pairs (x, ξ), where x indicates a location in the base space and ξ denotes the frequency aspect in the cotangent bundle, excluding the zero section to focus on non-trivial directions. Formally, it is expressed as:

$$\mathrm{WF}(u) = \{(x, \xi) \in T^*X \setminus 0 \mid u \text{ is not smooth at } x \text{ in direction } \xi\}. \tag{33.1}$$

The construction of the wavefront set involves the use of the Fourier transform $\mathcal{F}$, mapping a function or distribution into the frequency domain:

$$\mathcal{F}(u)(\xi) = \int_{\mathbb{R}^n} u(x) e^{-2\pi i x \cdot \xi} \, dx. \tag{33.2}$$

The decay of $\mathcal{F}(u)(\xi)$ in certain directions provides information suitable for determining elements in the wavefront set.

Propagation of Singularities

In quantum field theory (QFT), understanding how singularities travel through spacetime under the influence of field equations is crucial. Microlocal analysis equips us with powerful tools to track these singularities, using the Hamiltonian flow on the cotangent bundle.

1   Microlocal Propagation Theorem

The microlocal propagation theorem establishes that the wavefront set of a solution propagates along the null bicharacteristics of the principal symbol of the corresponding differential operator. Given a hyperbolic operator P, with principal symbol $p(x, \xi)$, the theorem states:

$$p(x, \xi) = 0 \quad \text{implies} \quad \text{WF}(Pu) \subset \text{WF}(u), \tag{33.3}$$

and the propagation of singularities occurs along the flow of the Hamiltonian vector field H_p associated with $p(x, \xi)$.

$$H_p = \frac{\partial p}{\partial \xi} \frac{\partial}{\partial x} - \frac{\partial p}{\partial x} \frac{\partial}{\partial \xi}. \tag{33.4}$$

2   Applications in Quantum Field Theory

In QFT, field equations, often modeled as linear PDEs, benefit from microlocal analysis to examine the behavior of singularities. For instance, consider a Klein-Gordon field equation:

$$(\Box + m^2)\phi = 0, \tag{33.5}$$

where $\Box$ denotes the d'Alembertian operator. The wavefront set of solutions ϕ reveals how initial singularities project through spacetime, impacting causal structures and scattering processes.

Algorithms for Identifying Singularities

The computational aspects of microlocal analysis in QFT are intricate and require algorithmic solutions to efficiently compute wavefront sets and follow the propagation paths of singularities.

Algorithm 15: Algorithm to Compute Approximate Wavefront Set

Input: A distribution u defined on a suitable domain
Output: The wavefront set $\text{WF}(u)$
Step 1: Choose a finite covering of X by coordinate charts.
Step 2: In each chart, compute the local Fourier transform $\mathcal{F}_\chi(u)$.
Step 3: Analyze the decay of $\mathcal{F}_\chi(u)$ in each direction.
Step 4: Identify (x, ξ) where $\mathcal{F}_\chi(u)$ shows non-trivial growth, hinting direction of singularities.
Return Wavefront set $\text{WF}(u)$ as a set of identified singular pairs (x, ξ).

---

1   Implementing Hamiltonian Flow

Analyzing the evolution of singularities via Hamiltonian flow provides crucial insights into the properties of solutions to field equations. By discretizing the Hamiltonian equations:

$$\frac{dx}{dt} = \frac{\partial p}{\partial \xi}, \tag{33.6}$$

$$\frac{d\xi}{dt} = -\frac{\partial p}{\partial x}, \tag{33.7}$$

numerical schemes enable tracking singularity propagation across the spacetime manifold.

Practice Problems 1

1. Define the wavefront set for a distribution u and explain how it provides insight into the singularities of u.

2. Consider a distribution u on $\mathbb{R}^n$. Describe how the Fourier transform $\mathcal{F}(u)(\xi)$ is used in determining the wavefront set of u.

3. Explain the microlocal propagation theorem and discuss its significance in the propagation of singularities for the Klein-Gordon equation.

4. Derive the Hamiltonian vector field H_p for a given principal symbol $p(x, \xi)$ and demonstrate its role in the propagation of singularities.

5. Illustrate with an example how microlocal analysis can be utilized to study singularities in a solution to a PDE commonly encountered in quantum field theory.

6. Outline an algorithm to numerically compute the wavefront set for a simple distribution and discuss the potential challenges in its implementation.

Answers 1

1. **Defining the Wavefront Set:**

   The wavefront set $\mathrm{WF}(u)$ for a distribution u is a collection of points (x, ξ) in the cotangent space $T^*X \setminus 0$, where u fails to be smooth at x in the direction of ξ. This set captures both the position and the directional information of singularities, offering a more geometric understanding of the singular structures within a distribution compared to general singular support. By excluding the zero section, it focuses on non-trivial directions where the distribution's Fourier transform does not decay rapidly.

2. **Fourier Transform in Wavefront Set Determination:**

   The Fourier transform $\mathcal{F}(u)(\xi)$ of a distribution u provides essential information about the distribution's behavior in the frequency domain. The decay properties of $\mathcal{F}(u)(\xi)$ with respect to ξ are central to characterizing the wavefront set. Specifically, rapid decay in ξ implies smoothness in corresponding spatial domain directions. Consequently, by examining directions where the decay is insufficient, one identifies potential singularity directions contributing to the wavefront set.

3. **Microlocal Propagation Theorem:**

   The microlocal propagation theorem states that the wavefront set of a solution to a differential equation propagates along the null bicharacteristics defined by the principal symbol of the operator. For a hyperbolic equation like the Klein-Gordon equation $(\Box + m^2)\phi = 0$, the theorem implies the wavefront set of ϕ propagates along the characteristics derived from the equation's hyperbolic nature. This theorem is pivotal as it helps in understanding how initial singularities evolve over time, giving insights into the causal structure and influence of field operators.

4. **Deriving the Hamiltonian Vector Field:**

   Given a principal symbol $p(x, \xi)$, the Hamiltonian vector field H_p is defined by:

   $$H_p = \frac{\partial p}{\partial \xi} \frac{\partial}{\partial x} - \frac{\partial p}{\partial x} \frac{\partial}{\partial \xi}$$

   This field captures the flow along which singularities move within the phase space. In the microlocal context, H_p determines how wavefront sets evolve, thereby linking the geometry of solutions to their analytic properties.

5. **Example of Microlocal Analysis in QFT:**

   Consider solutions to the Klein-Gordon equation $(\Box + m^2)\phi = 0$. Microlocal analysis examines the impact of initial singularities of ϕ_0 on subsequent states $\phi(t, x)$. Through the precision in phase space behavior, one can determine transport and interaction of singularities across spacetime regions, critically affecting scattering theory and causal behavior analysis in quantum field discussions.

6. **Algorithm to Compute Wavefront Set:**

Algorithm 16: Algorithm for Computing Wavefront Set

Input: A distribution u
Output: The wavefront set $\mathrm{WF}(u)$
Step 1: Cover the base space X with finite coordinate patches.
Step 2: Compute local Fourier transforms $\mathcal{F}_\chi(u)$ on each patch.
Step 3: Analyze the directional decay of $\mathcal{F}_\chi(u)$.
Step 4: Identify non-decay directions (x, ξ), indicating wavefront set candidates.
Return $\mathrm{WF}(u)$ as a culmination of identified pairs.
Challenges: Directional decay insights necessitate high-resolution Fourier analysis. Handling computational complexities and ensuring non-artificial truncation-induced artifacts in singularity identification remains intricate yet feasible.

Practice Problems 2

1. Define the concept of a wavefront set $\mathrm{WF}(u)$ for a distribution u and describe its significance in the study of singularities.

2. For a given distribution u, express the wavefront set $\mathrm{WF}(u)$ in terms of the behavior of its Fourier transform. Discuss the role of the cotangent space in this definition.

3. Explain the Microlocal Propagation Theorem and its implications for the propagation of singularities in solutions to partial differential equations.

4. Describe the Hamiltonian flow on the cotangent bundle and its relevance in determining the trajectory of singularities according to the Microlocal Propagation Theorem.

5. Consider the Klein-Gordon field equation. Explain how microlocal analysis can assist in understanding the wavefront set of its solutions. Provide a detailed example.

6. Outline an algorithmic approach to numerically compute the wavefront set of a distribution using its Fourier transform.

Answers 2

1. **Define the concept of a wavefront set WF(u) for a distribution u and describe its significance in the study of singularities.**

   Solution: The wavefront set WF(u) of a distribution u is a subset of the cotangent space $T^*X \setminus 0$ consisting of pairs (x, ξ), where x is a point in the base space and ξ represents a non-zero covector. The wavefront set identifies the directions ξ at which the distribution u fails to be smooth at the point x. By excluding the zero section, the wavefront set focuses on how discontinuities are oriented in space. This concept is significant because it provides comprehensive information about singularities, including their locations and directions, which is crucial for understanding the behavior of solutions to partial differential equations (PDEs), particularly in quantum field theory.

2. **For a given distribution u, express the wavefront set WF(u) in terms of the behavior of its Fourier transform. Discuss the role of the cotangent space in this definition.**

   Solution: The wavefront set WF(u) is determined by the asymptotic behavior of the Fourier transform $\mathcal{F}(u)(\xi)$ as ξ varies. Specifically, a point $(x, \xi) \in T^*X \setminus 0$ is in WF(u) if for any smooth function χ supported near x, the Fourier transform $\mathcal{F}(\chi u)(\xi)$ fails to exhibit rapid decay in the direction of ξ. The cotangent space T^*X is crucial in this setup because it provides the geometric setting for pairing spatial locations x with directions ξ, offering a full phase-space description for the analysis of singularities.

3. **Explain the Microlocal Propagation Theorem and its implications for the propagation of singularities in solutions to partial differential equations.**

   Solution: The Microlocal Propagation Theorem states that the singularities of a solution to a PDE propagate along the characteristic curves defined by the Hamiltonian associated with the principal symbol of the differential operator. For a linear differential operator P with principal symbol $p(x, \xi)$, the theorem asserts that any singularity present in the wavefront set of a solution u moves along the null bicharacteristics of $p(x, \xi)$, which coincide with the integral curves of the Hamiltonian vector field H_p. This result is significant for predicting how the wavefront set WF(u) evolves, thus informing the understanding of the dynamics of singularities in solutions.

4. **Describe the Hamiltonian flow on the cotangent bundle and its relevance in determining the trajectory of singularities according to the Microlocal Propagation Theorem.**

   Solution: The Hamiltonian flow on the cotangent bundle arises from the Hamiltonian vector field H_p, defined by a principal symbol $p(x, \xi)$. The flow is given by the system of differential equations derived from Hamilton's equations:

   $$\frac{dx}{dt} = \frac{\partial p}{\partial \xi}, \quad \frac{d\xi}{dt} = -\frac{\partial p}{\partial x}.$$

   In the context of the Microlocal Propagation Theorem, the Hamiltonian flow dictates the paths along which singularities propagate in phase space. It provides a mechanism to trace the evolution of the wavefront set by describing how points (x, ξ) in $\mathrm{WF}(u)$ move, thus helping to predict how singular structures develop and interact over time.

5. **Consider the Klein-Gordon field equation. Explain how microlocal analysis can assist in understanding the wavefront set of its solutions. Provide a detailed example.**

   Solution: The Klein-Gordon equation, given by

   $$(\Box + m^2)\phi = 0,$$

   involves the d'Alembertian $\Box$ operator. Microlocal analysis can be used to determine the wavefront set $\mathrm{WF}(\phi)$ of its solutions by examining how singularities generated by initial data propagate along characteristics. For instance, suppose the initial data ϕ_0 has a singularity at some point with a specific direction ξ_0. Using microlocal methods, we track this singularity by solving Hamilton's equations with the principal symbol of $\Box + m^2$, which in a simple flat spacetime translates to following null bicharacteristics of the d'Alembertian. These propagating characteristics will describe how the initial singularity evolves, crucially affecting the causal structure perceived from the solutions.

6. **Outline an algorithmic approach to numerically compute the wavefront set of a distribution using its Fourier transform.**

   Solution: An algorithm to numerically compute the wavefront set $\mathrm{WF}(u)$ involves the following steps:

   Algorithm 17: Algorithm to Compute Approximate Wavefront Set
   ---
   Step 1: Cover the domain X with finite coordinate charts.
   Step 2: In each chart, compute the local Fourier transform $\mathcal{F}_\chi(u)$ using FFT (Fast Fourier Transform) techniques.
   Step 3: Analyze the decay behavior of $\mathcal{F}_\chi(u)$ in different directions ξ.
   Step 4: Identify locations (x, ξ) where rapid decay fails, indicating potential singularity directions. Apply thresholds for numerical stability.
   Return the set of (x, ξ) pairs as the computed wavefront set $\mathrm{WF}(u)$.

   This approach allows the numerical estimation of singularities for further study of their properties and evolution using microlocal methods.

Practice Problems 3

1. Define the wavefront set $\mathrm{WF}(u)$ for a given distribution u and discuss its significance in the analysis of singularities.

2. Explain the construction of the wavefront set using the Fourier transform. Provide a step-by-step elucidation of how $\mathcal{F}(u)(\xi)$ aids in identifying singularities.

3. Describe the role of the Hamiltonian vector field H_p in the propagation of singularities. How does it relate to the principal symbol of a differential operator?

4. For the Klein-Gordon field equation $(\Box + m^2)\phi = 0$, illustrate the usage of microlocal analysis to determine the propagation of singularities in the context of quantum field theory.

5. Discuss the computational challenges involved in identifying wavefront sets and propagating singularities in quantum field theory. How can algorithms assist in this process?

6. Develop a simple algorithm to numerically simulate the propagation of singularities based on the Hamiltonian flow equations. Explain each step in the algorithm.

Answers 3

1. Define the wavefront set $WF(u)$ for a given distribution u and discuss its significance in the analysis of singularities.

Solution:
The wavefront set $\text{WF}(u)$ is a crucial concept in microlocal analysis that describes the set of singularities of a distribution in both space and frequency directions. Formally, it is the set of points (x, ξ) in the cotangent space $T^*X \setminus 0$ such that u exhibits a lack of smoothness at x in the direction ξ. This allows one to capture the precise nature and directionality of singularities, providing refined information beyond that offered by traditional pointwise analysis. The significance lies in its ability to help identify where singular behavior occurs and in which directions it might propagate, thus allowing a deeper understanding of the behavior of solutions to PDEs and their implications in physics.

2. Explain the construction of the wavefront set using the Fourier transform. Provide a step-by-step elucidation of how $\mathcal{F}(u)(\xi)$ aids in identifying singularities.
   Solution:
   To construct the wavefront set of a distribution u, follow these steps:

   (a) Perform a local Fourier transform on u, denoted by $\mathcal{F}_\chi(u)$, where χ is a smooth cutoff function, effectively making u compactly supported.

   (b) Calculate $\mathcal{F}_\chi(u)(\xi) = \int e^{-2\pi i x \cdot \xi} u(x) \chi(x) \, dx$.

   (c) Analyze the decay of $\mathcal{F}_\chi(u)(\xi)$ as $\|\xi\| \to \infty$. If $\mathcal{F}_\chi(u)$ decays rapidly in a direction ξ, then u is smooth in that direction.

   (d) Identify the pairs (x, ξ) for which $\mathcal{F}_\chi(u)(\xi)$ does not decay, indicating the presence of singularities.

   By understanding which directions ξ maintain non-trivial growth, the wavefront set $\text{WF}(u)$ is constructed, pinpointing where and how the distribution exhibits singular behavior.

3. Describe the role of the Hamiltonian vector field H_p in the propagation of singularities. How does it relate to the principal symbol of a differential operator?
   Solution:
   The Hamiltonian vector field H_p is essential for describing the evolution of singularities across the cotangent bundle in phase space. For a differential operator P with principal symbol $p(x, \xi)$, H_p is given by:

   $$H_p = \frac{\partial p}{\partial \xi} \frac{\partial}{\partial x} - \frac{\partial p}{\partial x} \frac{\partial}{\partial \xi}.$$

   Singularities propagate along the integral curves of H_p, which are the "characteristics" or "bicharacteristics" determined by the equation $p(x, \xi) = 0$. The propagation theorem states that singularities move along these null bicharacteristics, emphasizing the importance of H_p, as it governs the trajectory of these points in phase space.

4. For the Klein-Gordon field equation $(\Box + m^2)\phi = 0$, illustrate the usage of microlocal analysis to determine the propagation of singularities in the context of quantum field theory.
   Solution:
   Consider the Klein-Gordon equation $(\Box + m^2)\phi = 0$, where $\Box$ is the d'Alembertian operator. The principal symbol $p(x, \xi)$ is derived from $\Box + m^2$, with focus specifically on the d'Alembertian term. Applying microlocal analysis:

   (a) Compute the Hamiltonian vector field H_p, which describes the null bicharacteristics along which singularities travel.

   (b) Analyze how an initial point in the wavefront set $\text{WF}(\phi)$ propagates, by solving the equations:

   $$\frac{dx}{dt} = \frac{\partial p}{\partial \xi}, \quad \frac{d\xi}{dt} = -\frac{\partial p}{\partial x}.$$

   (c) Understand the trajectory of singularities through spacetime, providing insights into causal structures and potential scattering effects in quantum field theory.

   This methodology reveals the detailed movement of singularities, offering a rigorous basis for understanding physical phenomena such as causality and interactions.

5. Discuss the computational challenges involved in identifying wavefront sets and propagating singularities in quantum field theory. How can algorithms assist in this process?

Solution:

Computational challenges in wavefront set identification and singularity propagation arise from the necessity of accurately calculating local Fourier transforms, handling the large amount of data involved, and numerically solving Hamiltonian flow equations. Algorithms can alleviate these challenges by:

(a) Designing efficient numerical schemes for computing local Fourier transforms on manageable domains, ensuring accurate decay assessments.

(b) Implementing grid-based or adaptive methods to capture variations in singular behavior, allowing a computational focus on regions of interest.

(c) Developing robust time-stepping algorithms for solving Hamiltonian equations, facilitating the tracking of singularity paths over time.

Such algorithms streamline the analysis, making it feasible to explore large-scale problems typical in quantum field theory.

6. Develop a simple algorithm to numerically simulate the propagation of singularities based on the Hamiltonian flow equations. Explain each step in the algorithm.

Solution:

To numerically simulate singularity propagation, consider the basic steps:

Algorithm 18: Algorithm to Simulate Singularity Propagation

Input: Initial points (x_0, ξ_0) in WF(u), duration T, time step Δt
Output: Trajectories of singularities $(x(t), \xi(t))$
Initialize: Set $(x, \xi) = (x_0, \xi_0)$.
for *each time t from 0 to T with step Δt* **do**

> **Step 1:** Calculate derivatives $\frac{\partial p}{\partial \xi}(x, \xi)$ and $\frac{\partial p}{\partial x}(x, \xi)$.
> **Step 2:** Update positions using
>
> $$x(t + \Delta t) = x(t) + \Delta t \cdot \frac{\partial p}{\partial \xi},$$
>
> $$\xi(t + \Delta t) = \xi(t) - \Delta t \cdot \frac{\partial p}{\partial x}.$$
>
> **Step 3:** Record the updated values of (x, ξ).

Return: The series of updated positions (x, ξ).

This algorithm effectively maps the trajectories of singularities, enabling an empirical exploration of how they propagate through the spacetime domain according to the Hamiltonian dynamics.

Made in United States
Troutdale, OR
05/04/2025

31090363R00184